Thomas Haight Leggett

Electric Power Transmission Plants

And the Use of Electricity in Mining Operations

Thomas Haight Leggett

Electric Power Transmission Plants
And the Use of Electricity in Mining Operations

ISBN/EAN: 9783337027636

Printed in Europe, USA, Canada, Australia, Japan

Cover: Foto ©berggeist007 / pixelio.de

More available books at **www.hansebooks.com**

ELECTRIC POWER TRANSMISSION PLANTS

AND THE

USE OF ELECTRICITY IN MINING OPERATIONS.

By THOMAS HAIGHT LEGGETT,
Bodie, Mono County, California.

———

[WRITTEN FOR THE TWELFTH REPORT OF THE STATE MINERALOGIST, 1894.]

SACRAMENTO:
STATE OFFICE, : : : A. J. JOHNSTON, SUPT. STATE PRINTING.
1894.

ELECTRIC POWER TRANSMISSION PLANTS AND THE USE OF ELECTRICITY IN MINING OPERATIONS.

By Thomas Haight Leggett, of Bodie, Mono County, California.

Some one has aptly spoken of California as the Switzerland of America. Certainly the rugged scenery of its snow-capped Sierra, and its numerous lakes and mountain streams, justify, in part, the simile. In Switzerland they have been quick to realize the advantages to be derived from the utilization of their water powers for the generation of electric power, and its transmission to distant points; here, in California, we are but beginning to grasp the situation.

In electricity the miner has undoubtedly gained a most efficient and valuable ally. Through its aid the latent power of the many streams now running idly down the mountain slopes can be made available, and brought across long stretches of country by means of a simple line of wire, to operate the machinery of mine and mill.

In sections where no water powers are available, and fuel is scarce and dear, electricity may be generated at the center of fuel supply, and the power transmitted from this central station to operate a number of mills and hoisting works in the distant mining camp. One of the great advantages of electric power is its adaptability to ready subdivision into small units without material loss of power, by reason of the high efficiency now developed by the best types of dynamos. Hence, separate motors may be used in the mill for running crushers, stamps, concentrators, pans, etc., or in the mine for hoisting-engines, pumps, and air compressors, effecting a very appreciable saving when any of these machines are idle. To accomplish this requires the use of the direct current, but this can be readily obtained from the alternating where such is used for the transmission, by employing rotary transformers, or "motor generators," of high efficiency.

In a letter to the writer, accompanying photographs illustrating the Telluride, Colo., transmission plant, hereinafter described, Mr. Chas. F. Scott gives the following excellent résumé of the present status of electricity in the field of mining:

"In the introduction of electrical apparatus to the operations of the mining industries of the West, the field of electrical power transmission is extending upon lines which have already been well established in other industries. The electric motor is becoming an important factor in almost every industry in which power is utilized. One of the most notable instances is in electric traction. The electric street railway motor has not only almost entirely replaced animal power, but it has wonderfully increased the speed, comfort, and economy of street railway operation, and has also extended it to distances and classes of service which were previously impracticable. The early railway motor had many and peculiar difficulties to overcome, but the problems incident to it have been rapidly surmounted.

"Results similar to those which have been attained in street railway working are to be anticipated in the application of the electric motor to the mining industry. Not only will the work which is now performed be done in many cases with increased ease and economy, but the introduction of the motor will lead to new methods of operation. Mining possesses many difficult and peculiar requirements for the application of power. The motor, on the other hand, possesses characteristics which render it capable of being adapted to a very great variety of conditions.

"The work which has already been accomplished in the new plants which have been installed, promises much for the future. The first work has been under difficulties which are incident to every new undertaking. The principal difficulties which have manifested themselves are, however, not fundamental ones; they are principally due to mechanical difficulties which are more or less readily recognized, and usually indicate a ready method of solution. A second trouble, which has promised at times to be very serious, is the effects resulting from the atmospheric conditions in the mining country. Lightning in many places has been extremely severe, and methods of protection were required which were impossible to devise before the conditions had been learned from experience. The necessity for protection has been followed by the means of protection, and electrical installations need no longer be in peril from lightning, if properly protected.

"The experience and progress which have come from other applications of electricity can be taken advantage of in application to mining work. The constant improvements and advances which are being made in the manufacture of electrical apparatus make it possible to secure at the present time apparatus which is better adapted for its work than could have been secured a few years ago.

"There is often an apprehension, on the part of those who are not familiar with electrical apparatus, that it is a fundamental failure if it does not at once begin and continue in satisfactory operation. Those, however, who are acquainted with electrical machinery, and who understand the nature of the difficulties which develop, may readily see that the fundamental elements in electrical power transmission and distribution are not involved in these difficulties, but that they arise from incidental features which can be readily corrected. The work which has already been accomplished shows the possibilities which are open in the field of electrical mining, and promises much for the future."

The transmission of 100 horse-power a distance of 109 miles, from Frankfort to Lauffen, Germany, in 1891, showed conclusively that from an engineering standpoint, at least, the transmission of power over long distances by electricity was perfectly practicable; though in this particular instance it was not a commercial success, nor was it intended to be, since the power was used for exhibition purposes only. Since then, however, plants have been installed both in Europe and in the United States, and are to-day successfully transmitting electricity for lighting and power purposes over distances ranging from 1 to 30 miles.

It will be proper to outline here the various methods of transmitting power over long distances by electricity, but for full information on this subject recourse must be had to the technical writers in the electrical journals and society transactions.*

* See W. F. C. Hasson's paper on "Electric Transmission of Power Long Distances," Transactions of the Technical Society of the Pacific Coast, Vol. X, No. 4: "Long Distance Transmission for Lighting and Power," by Chas. F. Scott, E.E., Vol. IX of Transactions of American Institute of Electrical Engineers; also pamphlet on Long Distance Transmission by L. B. Stillwell, E.E., issued by Westinghouse Electrical and Manufacturing Co., Pittsburg, Pa.

Power may be transmitted by means of electricity by—
1st. The direct or continuous current.

$$
\left.
\begin{array}{l}
\text{2d. The alternating current --}
\left\{
\begin{array}{l}
(a)\text{Single phase, 2-wire synchronous} \\
\quad\text{system.} \\
(b)\text{Two-phase, 4-wire} \\
\quad\text{system.} \\
(c)\text{Polyphase system;} \\
\quad\text{usually 3-phase} \\
\quad\text{with 3 wires.}
\end{array}
\right.
\end{array}
\right\}
\begin{array}{l}
\text{Either synchro-} \\
\text{nous or inde-} \\
\text{pendent speeds} \\
\text{of generator} \\
\text{and motor, as} \\
\text{desired.}
\end{array}
$$

The direct or continuous current has the advantage that the motors are self-starting, and at practically full torque, or turning effect. The motor speed is quite independent of that of the generator, though this is not necessarily an advantage, inasmuch as, in synchronous systems, the governing of the generator speed regulates that of the motor as well, and therefore attention to the speed of but one machine is all that is required.

Direct-current dynamos labor under the disadvantage of working under comparatively low potentials, since they require a commutator to change the alternating current they generate into a continuous one, i. e., a current flowing constantly in one direction; and thus far it has been found impracticable to insulate this commutator for very high tensions. While it is asserted that* "direct-current machines of 5,000 volts are in regular and successful use for arc-lighting," it must be borne in mind that the requirements for furnishing light a limited number of hours each day are very different from the demands made upon electrical machines by a stamp mill or hoisting works, which require unintermittent operation, oftentimes including Sundays.

Hence, such a high-potential, direct-current machine, if in good running order to-day, would hardly be serviceable for long-distance transmission, and indeed the staunchest advocates of the direct current in this country have never installed such a machine for this purpose.

On the contrary, in several cases where electrical companies known to favor the direct-current system have had contracts for the installation of long-distance transmission plants, they have not attempted such, but have instead used an alternating-current system in every case where the distance exceeded three miles.

It is safe, therefore, to conclude that until these difficulties of commutator insulation are overcome,† this distance is the practical limit for direct-current transmission, unless a series arrangement of generators and motors be resorted to.

A low potential necessarily limits the distance of transmission, since the size of wire is directly proportional to the number of amperes of current to be carried; and since amperes times volts equals watts, of which 746 are equivalent to 1 horse-power, it follows that to transmit

---

* The "Electric Transmission of Power," Engineering Magazine, June, 1894, p. 393.

† Mr. E. H. Booth, in an article entitled "Electricity as applied to Mining Operations," published in "Industry" for June, 1892, says: "It is, however, a matter of difficulty to make commutators for potentials over 2,000 volts for direct-current generators, on account of the great number of segments required, and the difficulty of their proper insulation. While this voltage will be efficient and economical, both as regards cost of installation and of operation in many cases, conditions will also be met with requiring much higher voltages, which are at present commercially practicable only through the use of alternating currents."

100 horse-power, or 74,600 watts, a given distance at a pressure of 500 volts (the ordinary voltage of a direct-current dynamo), would require a current of 149.2 amperes, or a wire six times as large (sectional area six times as great) as that needed to deliver the same amount of power over the same distance at an electrical tension of 3,000 volts (25 amperes $\times$ 3,000 volts = 75,000 watts = 100 horse-power).

The series arrangement of generators and motors alluded to has been introduced abroad, notably in Switzerland, and brought there to a higher state of perfection than in this country. This application of the direct current for long-distance transmission requires a number of generators and an equal number of motors, making a complicated apparatus of excessive first cost, especially so since extra dynamos and motors must be provided; otherwise an accident to one machine disables the entire plant.

At Genoa, Italy, there is such a transmission at present in operation. The power transmitted is 300 horse-power over a distance of 18 miles. At the power stations, of which there are three, one below the other, there are four groups of dynamos, each group of two dynamos being driven by turbines (Piccard system) of 140 horse-power, working under heads varying from 225 to 495 meters.

These dynamos are connected in series, one group being held in reserve in case of accident to any of the others, and produce each a current of 47 amperes at 1,000 volts electrical tension, the resulting E. M. F. sometimes reaching 6,000 volts during the hours of maximum load. The motors are also connected in series, no one machine, it will be noted, carrying a potential exceeding 1,000 volts at any time. The power is utilized in operating a factory at the terminus of the line.

The lack of flexibility of the system and its inadaptability to a wide and varied range of work have often been spoken of by technical writers, and these disadvantages have prevented its successful competition with alternating-current systems for transmission—such as that from Niagara Falls to Buffalo, where the power is to be utilized for a great variety of work.

It has been cited as an advantage of the direct-current system that it is not liable to trouble from the static capacity and self-induction of the line occurring with the alternating-current. Self-induction will reduce the potential at the motor end of the line, while static capacity will act in the opposite direction and increase the E. M. F., thus tending to counteract the effect of self-induction.

The discussion of these technicalities can safely be left to the electricians, but the writer can state from experience with a transmission by the alternating-current synchronous system of 120 horse-power over a distance of 12½ miles that no trouble whatever has arisen from these causes. (See table showing the line-loss and the efficiency of this transmission.)

The three types of alternating-current machines, viz., the single-phase synchronous, the double-phase, and the three-phase generators, may, for purposes of comparison, be likened, respectively, to the single-cylinder steam engine, the double or two-cylinder engine with crank arms at 90°, and the three-cylinder engine with as many crank arms set at an angle of 120° each with the other; the electrical impulses bear just these relations with each other in the armature of the dynamo.

The single-phase generators and motors are necessarily synchronous, and the latter are not self-starting, but must be brought up to the generator speed before the line current can be led into its armature; while the polyphase machines are self-starting under light load, but not under full load.

It is evident that for hoisting and similar work, where full load is thrown on the machine at once, alternating-current motors do not possess the advantages of direct-current machines, which start readily under such conditions, and for short periods can be greatly overloaded without damage.

If, therefore, it be desired to use electric power in all departments of a mining plant, the electricity being generated at a considerable distance from the works, cheapness of first cost and of copper conductors can be obtained by using a high-potential alternating system, with raising and lowering transformers if necessary; while by using rotary transformers, or motor generators, as they are sometimes termed, at the delivery end of the line, direct current can be obtained for all work requiring self-starting motors. These machines used for transforming alternating current into direct current at various potentials have a common field, and two windings upon the armature revolving within it, one of which receives the alternating current and acts as a motor, while the other generates the required direct current.*

For long-distance transmission the alternating current possesses the great advantage of being convertible from a low to a high potential, or vice versa, by means of a simple transformer, without moving parts, thereby effecting a great saving in copper, since, as already shown, the greater the E. M. F. the less number of amperes of current required to transmit a given power, and hence the smaller wire demanded. Single-phase, alternating-current motors, while not self-starting, may be heavily overloaded without pulling them out of synchronism and causing them to stop; and should the latter occur, no damage will result under ordinary conditions, since the self-induction of the armature will hold back the current for several minutes. They may be also heavily overloaded immediately after synchronizing. The 120 horse-power motor in the mill of the Standard Consolidated Mining Company at Bodie, Cal., has started all twenty stamps while resting upon the cams, though this, of course, is not the ordinary way of taking up the load. It shows that the motors will take an abnormally heavy load at the outset without damage beyond a little extra sparking at the commutator.

The two-phase four-wire system is equally adapted for both lighting and power, and it is not necessarily synchronous, though the advantage of speed regulation previously referred to makes it advisable to so operate the generator and motor wherever possible. It will furnish power through motors of either the rotating-field type (i. e., rotating magnetism) or the polyphase; and by means of commutating devices

---

* Induction motors (three-phase) are now being built by the General Electric Company, and quarter-phase machines by the Westinghouse Electric and Manufacturing Company, which it is claimed are fully equal in every respect to direct-current machines. This is a development only to be expected in view of the fact that the best electricians in the country have been devoting their best energies to the attainment of this most desired result; and it will greatly simplify any quarter-phase or three-phase transmission plant where the power is to be used for the various classes of work required in mining.

it can be made to supply direct current for all power and lighting work if so desired, and at a very high efficiency of transformation.

It is therefore particularly well adapted to mining requirements, which, as stated, demand motors starting immediately and with full torque for certain classes of work.

"There are two especially prominent types of these machines. The first of these, the double machine, has two fields and two armatures, the latter mounted on the same shaft. Each armature delivers alternating current to a two-wire circuit, and these circuits taken together constitute the four-wire circuit of the generator; or they may be so connected as to constitute a three-wire circuit.

"Machines of the second type have single armatures with two windings, or with a single winding so connected to the ring collectors as to deliver two currents differing in their time relation or phase."*

Twelve machines of the first type and of 1,000 horse-power capacity were used by the Westinghouse Electric and Manufacturing Company as single-phase generators for lighting purposes at the World's Fair. Some of these were used for power to run exhibit motors, and in these cases were connected as quarter-phase (two-phase) machines.

There is a decided advantage in this system over the three-phase in the distribution of load on the two circuits of which it is composed, as the machine can be designed to regulate each current independently, i. e., maintain a constant E. M. F. with varying loads on the circuit, which cannot be done with the three-phase system.

This advantage largely offsets the saving in copper of the latter system, which saving can be put roughly at about 25 per cent over that of either the single two-wire or two-phase four-wire systems. These latter stand on about an equal footing as regards the amount of copper required for transmitting power over a given distance at a stated potential.

"In a paper read before Section 'G' of the British Association, on September 18, 1893, Mr. Gilbert Kapp makes the following statement: 'If we put all the systems on the same footing as regards efficiency and safety of insulation, we find the following, viz.: For the transmission of a certain power over a given distance * * * the single-phase alternating and the two-phase four-wire system will require 200 tons, the two-phase three-wire system will require 290 tons, and the three-phase three-wire system only 150 tons. As far as the line is concerned there is thus a distinct advantage in the employment of the three-phase system.' "†

For the amounts and cost of copper required for transmitting power over varying distances and under different potentials, the reader is referred to the papers by Messrs. Hasson and Stillwell, already cited.

From the foregoing it would appear that for the ordinary work of stamp mills, where single large units of power are chiefly needed, the single-phase synchronous motors are well adapted to meet all requirements where the power is transmitted from a distance too great for the use of the direct current; while for a more extended and varied use of

---

*From "Transmission of Power," a pamphlet issued by Westinghouse E. & M. Co., and prepared by L. B. Stillwell, E.E.
† "The Electrical Engineer" (N. Y.), January 17, 1894, p. 42.

such power the polyphase systems are more economical and compre-
hensive, more especially the two-phase four-wire method.*

In the paper by Mr. E. H. Booth, already referred to, he speaks of
the use of separate motors for each stamp battery, and for groups of
four pans and two settlers each, thus doing away with heavy and
expensive line-shafting, belt alley-way, etc.

Such an extreme subdivision of the power, however, would result in
a heavy loss of efficiency, and is further highly impracticable at present
on account of the high speeds at which electric machines operate, neces-
sitating counter shafting to reduce the revolutions to the slow speed of
pan and cam shafts.

It is better, therefore, for milling work, to use a single large motor to
operate the stamps, pans, etc., with perhaps one or two small ones for
rock-crushers and concentrators in cases where the cost of the power or
the production of higher-grade concentrates makes this an object.

The following description of the power-transmission plant of the
Standard Consolidated Mining Company, at Bodie, Cal., is taken from
the writer's paper read before the American Institute of Mining Engineers
at the Virginia Beach meeting, February, 1894:

THE ELECTRIC POWER TRANSMISSION PLANT OF THE STANDARD CONSOLI-
DATED MINING COMPANY.

At Bodie, Mono County, Cal., the ruling price for wood has been, for
years past, $10 per cord, so that the monthly fuel bills of a 20-stamp mill,
crushing and amalgamating 50 tons of ore per day, would often amount to
$2,000. To reduce this excessive cost of motive power was the problem in
hand, and the use of electricity generated by water power has solved it.
No sufficient water power could be found nearer than 12¼ miles, the distance from Bodie in a straight line over the hills to the east flank of the
Sierra Nevada. This distance is just at that intermediate point where
the cost of transformers about equals the difference in cost between a
No. 1 and a No. 6 copper wire (it is not advisable to use any lighter
wire than No. 6, on account of its liability to rupture during storms).
Hence it was deemed better not to use converters, since they would only
complicate the apparatus, without effecting a saving in cost.

Water-Power Plant.

An excellent water power was found in a mountain stream on the
north slope of Castle Peak, in the Sierra Nevada, known as Green Creek,
and forming one of the chief sources of the East Walker River. This
stream carries 400 in. of water during the dry season; and ten times that
amount during the time of melting snows.

An old ditch was cleared out and rebuilt for a length of 4,570 ft., and
a site was selected for a power-house 355 ft. vertically below its lower

---

* For a full comparison of the relative advantages of the two-phase and the three-
phase circuits, see " Polyphase Transmission," by Chas. F. Scott, in the " Electrical Engineer " (N. Y.), March 21, 1894. In this article Mr. Scott proposes a combination of the
two-phase and three-phase systems, generating under the first system, and by means of
special transformers (while also raising the potential if so desired) changing to the
three-phase for the transmitting line and again converting to the two-phase current at
the delivery end of the transmission, thereby uniting the advantages of saving in copper,
of the one system, with those of greater simplicity, less cost of apparatus, and better
regulation of the other method (the two-phase).

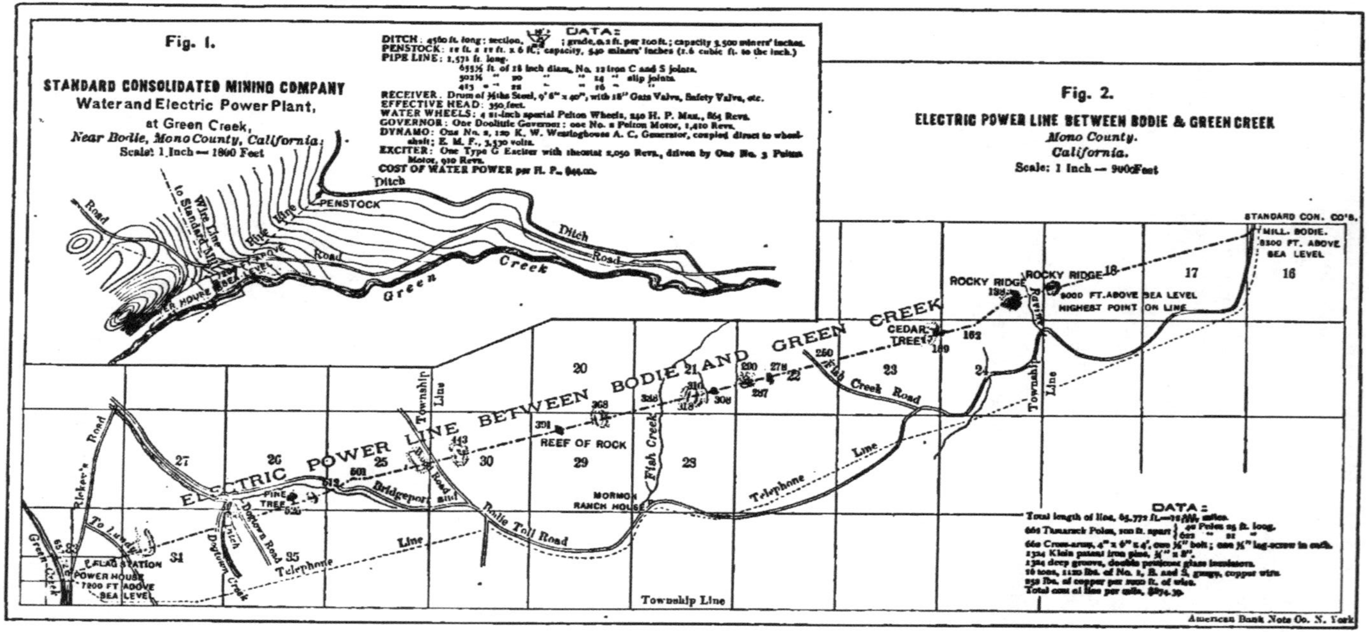

Fig. I.

STANDARD CONSOLIDATED MINING COMPANY
Water and Electric Power Plant,
at Green Creek,
Near Bodie, Mono County, California;

Scale: 1 Inch — 1960 Feet

DATA:
DITCH, 4380 ft. long; section, ⌷
PENSTOCK, 11 ft. 1 rft. x 6 ft., capacity, 3,500 miners' inches
PIPE LINE, 2,931 ft. long,
    2936 ft. of 18 inch diam. No. 12 to 14 C. and S. joints
    ...
RECEIVER, Drum of 16ft. Head, 1'8" x 4'6", with 18" Gate Valve, Safety Valve, etc.
EFFECTIVE HEAD, 395 feet.
WATER WHEELS, 3 single Impulse Pelton Wheels, 240 H. P. Max., 85 Revs.
GOVERNOR, One Doubble Governor; 2 new No. 1 Pelton Motor, 4,000 Revs.
DYNAMO, One No. 4, 120 K. W. Westinghouse A. C. Generator, connect direct to wheel,
    shaft; One Type G Exciter with shaft 4,090 Revs., driven by One No. 3 Pelton
    Motor, 950 Revs.
EXCITER, One Type G Exciter
COST OF WATER POWER per H. P. Max.

Fig. 2.

ELECTRIC POWER LINE BETWEEN BODIE & GREEN CREEK
Mono County,
California.

Scale: 1 Inch — 9000 Feet

DATA:
Total length of line, 52,777 ft.

STANDARD CON. CO'S.
MILL. BODIE.
8300 FT. ABOVE SEA LEVEL

ROCKY RIDGE
9000 FT. ABOVE SEA LEVEL
HIGHEST POINT ON LINE

FLAG STATION
POWER HOUSE
7250 FT. ABOVE SEA LEVEL

Plate 1. Generator and Waterwheels in operation.

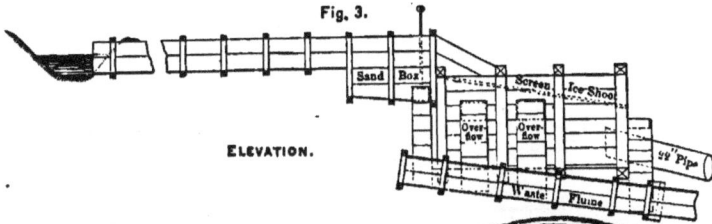

Fig. 3.

ELEVATION.

PENSTOCK AND FLUME.
Scale ¼in.= 3 ft.
1893.

Fig. 4.

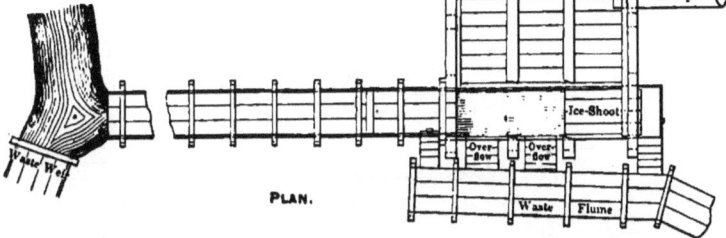

PLAN.

end. The ditch was made larger than necessary for power purposes alone, with the object of supplying other parties, when there was an excess of water.

The maps, Figs. 1 and 2, give the data with regard to the ditch and pipe; and Figs. 3 and 4 show the connecting flume, pressure-tank, and waste-weirs. The arrangement of the screen adopted, while it occasions a loss of head of a couple of feet, is greatly to be recommended where "anchor" and slush-ice form in a ditch during cold weather.

The pipe is of large diameter, in order to permit subsequent enlargement of the plant, and also to reduce loss of head by friction. It is fitted with three 2½ in. air valves, to prevent collapse in case of sudden rupture, and is anchored at proper intervals with straps of 1¼ in. round iron. The slip-joints extend to a vertical head of 220 ft., the remainder of the pipe being laid with collar-and-sleeve lead joints.

The pipe leads into a receiver 40 in. in diameter and 9½ ft. long, from which four taper-pipes lead the water, under pressure of 152 lbs. per square inch, to as many 21 in. Pelton waterwheels, each wheel being fitted with two nozzles and rated at 60 horse-power under the largest sized tips of 1⅛ in. diameter.

The speed of the wheels is 860 to 870 revolutions, and their shaft is connected by an insulated rigid coupling to the armature shaft of a 120 K.-W. A. C. generator. Plate I shows the generator and waterwheels in operation.

The accompanying plan (Fig. 5) shows the arrangement of the plant, one of the most interesting features of which is the water-governor formerly known as the "Doolittle," and now called the Pelton differential governor (Figs. 6 and 7). It operates butterfly-valves placed in the 5 in. pipes between the gate-valves and the diverging nozzles; and though this form of valve invariably "throttles" the water to a greater or less

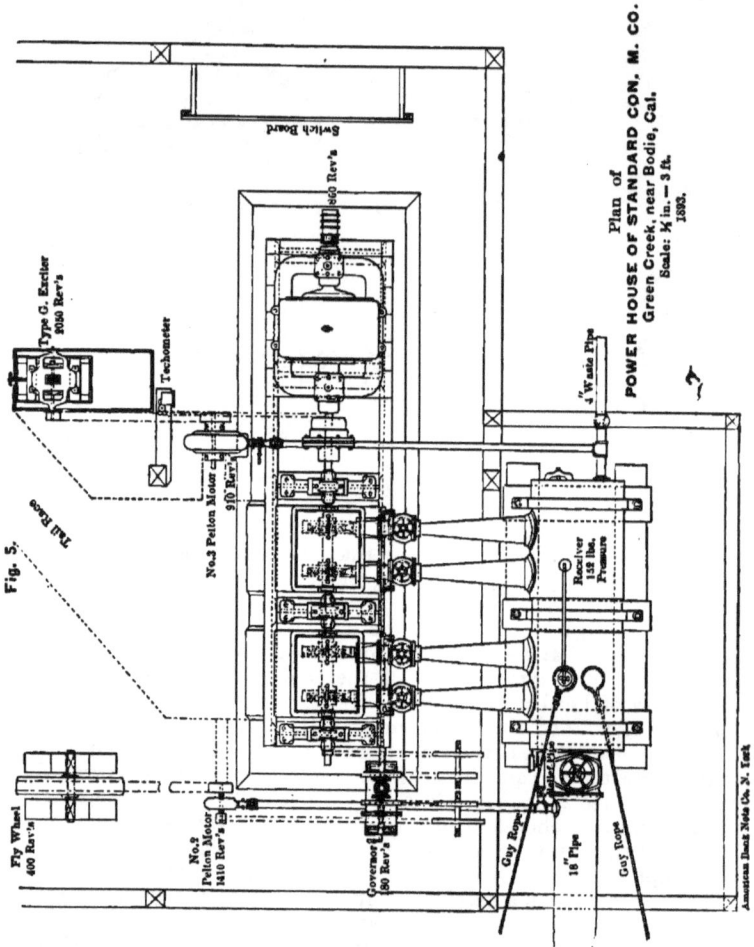

Fig. 5.

Plan of
POWER HOUSE OF STANDARD CON. M. CO.
Green Creek, near Bodie, Cal.
Scale: ¼ in. = 3 ft.
1893.

extent (according to the position of the valve), it is a most satisfactory way of controlling the power where the same is ample, and the loss due to this cause is of slight consequence. The governor operates as follows: Two 18 in. pulleys revolve loosely and in opposite directions on a shaft, one being driven from the waterwheel shaft and the other by a No. 2

Pelton motor. These pulleys have gears on their hubs which mesh into two other gear-wheels carried on an axis at right angles to the shaft and keyed fast to the latter. Beyond these wheels is a pinion, loose on the shaft and with ratchet-teeth cut in opposite directions on either side of its hub. Into these ratchet-teeth mesh corresponding circular ratchets, which are keyed to the shaft but free to move longitudinally along the same, and are thrown in or out of gear by a short lever and spring. The pinion engages a sector, which is fastened to the rod and levers that operate the butterfly-valves, and on the same rod is a hand-lever, by means of which the valves may also be opened or closed by simply throw-

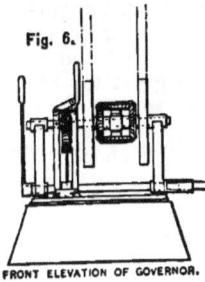

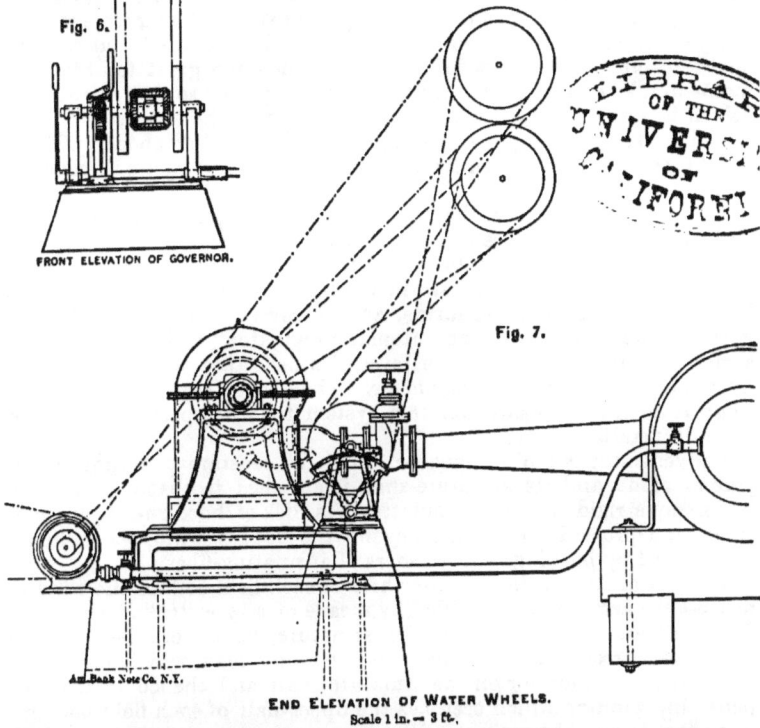

Fig. 6.

FRONT ELEVATION OF GOVERNOR.

Fig. 7.

Am. Bank Note Co. N.Y.

END ELEVATION OF WATER WHEELS.
Scale 1 in. = 3 ft.

ing out of mesh the circular ratchets alluded to and thereby detaching the governor. It is evident that when the two pulleys are revolving in opposite directions at exactly the same rate of speed, there will be no motion of the central gear-shaft, and none will be communicated to the pinion and sector and thence to the valves, to open or close them; while, on the other hand, a difference in speed of these pulleys will have the opposite effect. The belts driving them are therefore so arranged that a decrease in speed of the waterwheels will open the valves, and an increase will close them.

In starting up from rest, the governor is detached by throwing out the springs on the ratchets, and the valves are operated by the hand-lever.

After the wheels are at normal speed and the load is on, the ratchets are sprung into gear with the pinion, and the governor takes care of any and all variations, even to a complete throwing off of the load by pulling the main-current plug-switch at Bodie. The speed of the governor-pulleys, as first designed, was 60 revolutions. This was found to be too slow, and it was increased to 180 revolutions with most beneficial effects, developing a greater sensitiveness to small changes of load, and much quicker action, especially when all the load was thrown off at once. In the latter case, the increase in speed of the waterwheels did not at any time exceed 12 per cent before the governor began to close the valves.

It was further found necessary to furnish a constant resistance for the water motor that drives one side of the governor, to work against. In the original plan this was to be done by the exciter which furnishes current for the fields of the generator; but on trial it appeared that the load on the exciter was too variable, and at times too great for the little motor to take care of. The exciter was then placed so that it could be driven by either a larger size (No. 3) motor or by the waterwheel shaft-coupling (see plan of power-house); and a fly-wheel of about 1,500 lbs. weight was set to be driven by the smaller motor and insure its constant speed.

The great drawback to the use of water power for the generation of electricity has hitherto been the lack of a good water-governor, sufficiently sensitive and quick-acting to insure the vital factor of constant speed without bringing dangerous strain on the water pipe. In fact, in the Westinghouse plant at Telluride, and in several others of which the writer is aware, the "one-man automatic regulator" had to be used; i. e., a man sat with his hand on the lever of a deflecting nozzle and his eye fixed on a voltmeter or a techometer. The above-described governor is so great an improvement over this system that its operation has been given in detail.

The generator is a Westinghouse 120 K.-W. constant-potential twelve-pole machine, and its armature-shaft is attached to that of the water-wheels by a rigid coupling, insulated by a disk of hard rubber one inch thick, and projecting one inch beyond the flanges, while the bolts are surrounded by bushings and washers of insulating-fiber.

The initial current in the lower half of each field-coil, or the winding nearest the armature, is instilled by means of a type "G" D. C. exciter. The secondary winding, on the armature-spokes of the dynamo, generates current when the machine is under load, which is led to a twelve-bar commutator on the armature-shaft and thence to the compensating-winding which occupies the upper half of each field-bobbin.

As the load on the generator increases, more current flows through its armature-coils, and through a primary winding on the armature-spokes, thereby inducing, in the secondary winding, a heavier current, which, being led to the magnetic field as described, proportionately strengthens the same. When the generator is running without load, there being little or no current in its armature-coils, none is induced in the secondary winding, and the compensating-winding on the fields is without magnetic effect until the latter is required by work to be performed.

The potential of the generator under full load is 3,530 volts, but at present it is operating with about 3,390. The exciter carries a voltage of 105 to 112. A "D. C." voltmeter, recently placed on the switch-board to the left of the ground-detector and above the small rheostat, is in the

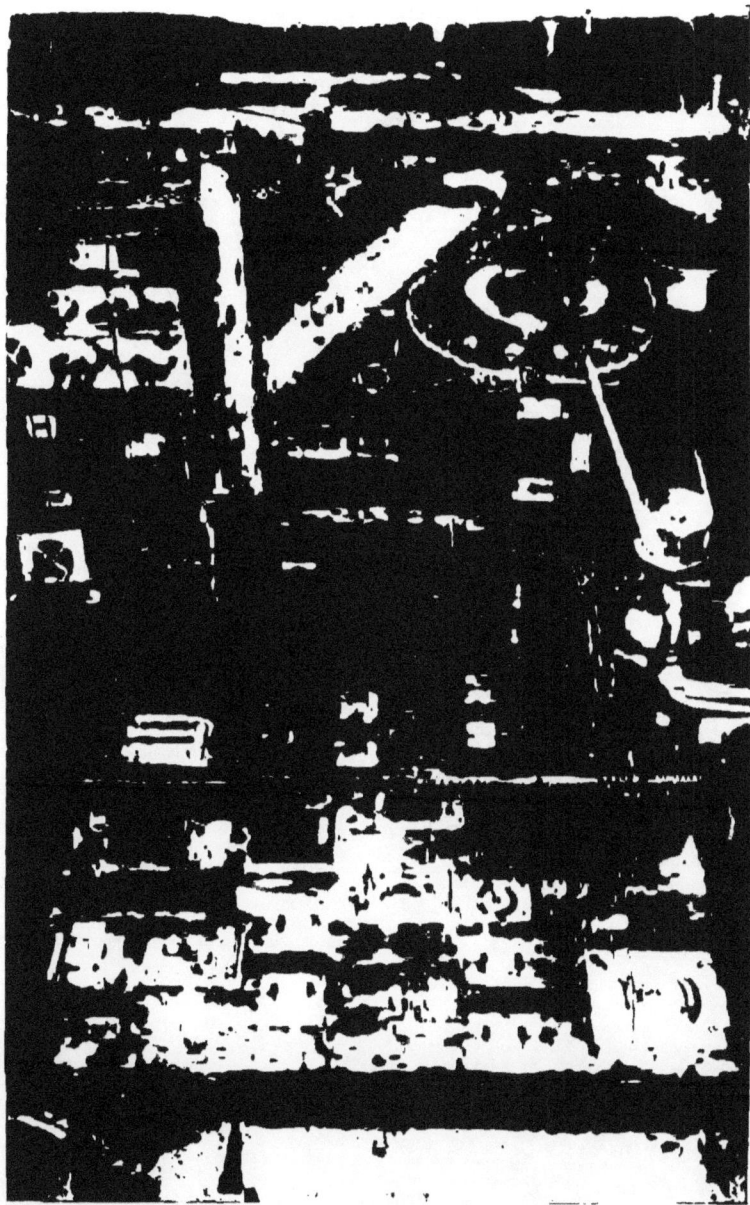

Plate II. Generator Switch-Board at Power-House. (Generator in operation; Exciter in foreground; Choke-Coils and Gap-Lightning-Arresters on separate board.).

main circuit of the exciter, recording the tension of its current and serving as a speed indicator when the machine is driven by the No. 3 motor. This is not necessary when driving from the wheel-shaft, as is sometimes done in winter, when pieces of ice give trouble in the small nozzle of the motor.

Plate II shows the generator switch-board at the power-house. The generator current is led from the collector-rings on the extreme end of the armature-shaft to the plug-sockets on the switch-board; and when the line-plugs are in these, the current follows the line to two similar sockets on the motor switch-board. The small converter in the upper middle of the switch-board has a transforming ratio of 30 to 1. Its primary coil is attached to the main-current wires from the generator, and its secondary to the A. C. voltmeter, immediately below it. A potential of 113 volts on the voltmeter is therefore equivalent to 3,390 on the dynamo current, which is the tension under normal load. The voltmeter does not, however, read 113 volts, but records 100 to 102 volts, the difference being due to the compensator (the instrument shown in the upper left-hand corner of the switch-board and connected with the voltmeter), the object being to reduce the reading by an amount about equal to the line-loss. This loss is estimated at 15 per cent under maximum load, and is but from 8 to 10 per cent under normal load, as will be shown later on.

The ammeter, and just below it the aluminum fuses, all of which are in the main circuit, are shown to the left of the voltmeter in the view of the generator switch-board.

Immediately to the left of the main-line plug-switches is the ground-detector with two lamps, one for each leg of the line, and each lamp with its converter behind it.

A press-button below the lamps makes the necessary connection with a ground wire. Without this connection made, the lamps show a red light on the filaments, due to the difference in potential of the two sides of the line; and should a "ground" occur on either leg of the wire-line, the corresponding lamp immediately burns at full candle power, while the other lamp proportionately diminishes.

The two-pole jaw-switch to the left of the switch-board is in the circuit from the exciter to the generator-fields, as are also the two fuses and the rheostat immediately below it. The small rheostat to the right of the fuse-blocks and the single-pole switch below it are in the shunt field-circuit of the exciter. By means of these two rheostats the potential of the generator is governed and the voltmeter kept at its proper reading, the large rheostat in the exciter and generator field-circuit permitting a quick regulation over a wide range, and the shunt-rheostat a finer and closer adjustment of the voltage.

When starting up the plant one attendant stands at the lever, controlling the admission of water to the wheels through the butterfly-valves, and the other at the switch-board, handling these two rheostats (most of the regulation is done by the large one), until the motor is in synchronism and at work, when the governor is thrown into gear, the voltage is finally adjusted, and the mechanism is then practically self-regulating for all ordinary changes of load. If, for instance, ten of the twenty stamps are to be hung up, or any or all of the eight continuous-pans in the mill are to be stopped, it is never necessary first to give word to the attendant at the power-house. The governor takes charge of such changes, even to the entire throwing off of the load, as before remarked. All the bearings of the generator and waterwheel shafts and of the exciter are self-oiling. The attendant has merely to keep on the *qui vive* and see that all is running smoothly. Any change in tone of the hum of the dynamo warns him at once of a change of conditions, the tone rising or falling according as the speed increases or diminishes, though ever so slightly.

To insure the all-important factor of constant speed, a techometer, registering to 1,200 revolutions, is belted to the waterwheel and dynamo shaft. Its dial faces the waterwheels, so that the attendant at the valve lever can readily maintain a uniform speed during the operation of "synchronizing" the motor and starting the mill, at which time the load is constantly varying.

In front of the jaw-switch on the switch-board there will be noticed, in the view of the latter, a steel spring, and also two cords attached to the handle of the large rheostat. These cords are led·around the side of the building to the attendant's place at the valve-lever, as is also the one that releases the catch of the spring. A pull on these cords opens the exciter main-circuit instantly, and puts in the entire resistance-box, thereby "killing" the fields of the generator and preventing any dangerous rise in electro-motive force, should the load be suddenly thrown off by a break in the wire-line, or other accident causing a sudden increase in the speed of the armature shaft. It should be explained that this arrangement was devised by the writer, before the speed of the governor was trebled, the constant-resistance fly-wheel was put in and other changes were made, giving more sensitive and perfect control of the water power; and it is left in place because it might still be of use in case of emergency. The power-house is lit by a small 10-light converter attached to the generator-circuit, and when the generator is not in operation, by current from the exciter. Plate III shows the power-house at Green Creek.

Wire-Line.

The length of the line is 67,760 ft., or 12.46 miles. The poles are of round tamarack timber, 21 ft. long, 6 in. in diameter at the top, set 4 ft. in the ground; poles 25 ft. long being used through the town, and along the line wherever there is danger of deep snowdrifts. They are placed 100 ft. apart, and fitted each with a 4 by 6 in. cross-arm, boxed into the pole, and held by one bolt and one lag-screw. The accompanying sketch (Fig. 8) shows the detail. The object of chamfering the ends of the cross-arms is to leave less room for the lodging of snow under the insulator.

The line crosses extremely rough country, not 500 yds. of which is level beyond the town limits. Most of the ground is very rocky, over 500 lbs. of dynamite being used in blasting the pole-holes. Plates IV and V are views along the line in summer.

The wire is of No. 1 (B. & S.) gauge, soft-drawn bare copper, and is attached to standard, double-petticoat, deep-groove glass insulators (Fig. 10) carried on Klein patent iron pins (Fig. 9). The distance between the wires is 3 ft. 8 in., and there are over 16.5 tons of copper in the line. The only objection found to the iron pins is their liability to be withdrawn from the cross-arm during a gale of wind, whenever there is an upward pull on the wire. To obviate this a number of pins were drilled

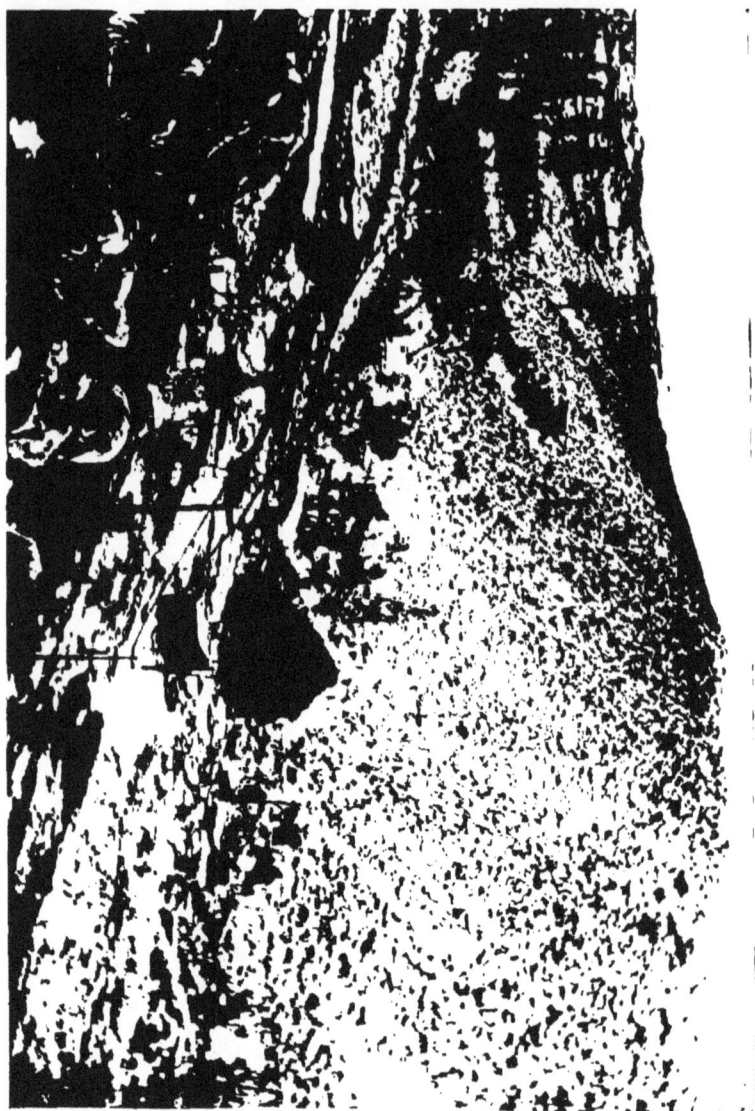

Plate III. Power-House at Green Creek.

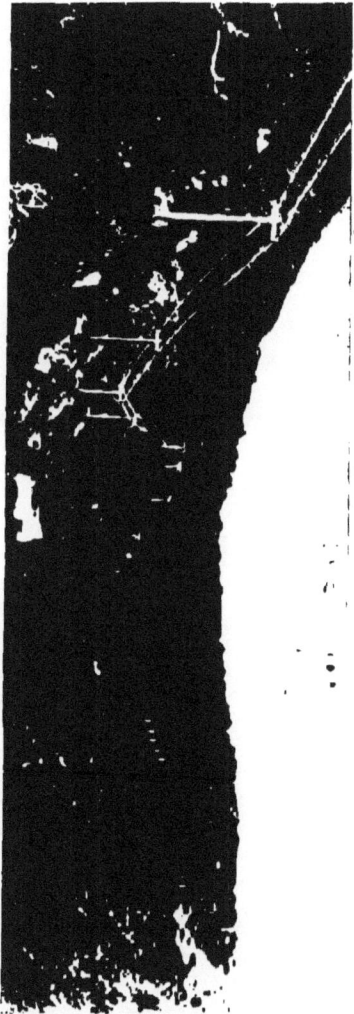

Plate IV. Summer View on Pole-Line, looking east, 10 miles from Bodie.

Plate V. Summer View on Pole-Line, looking west, 10 miles from Bodie.

Fig. 8.

8' 8" between Wires

6"

3' 6"

6"

8"

t. ? Clearance

3¼"

4"

16 ½"

4'

8' to 9"

Fig. 9.

1¼"

2"

8"

¾"

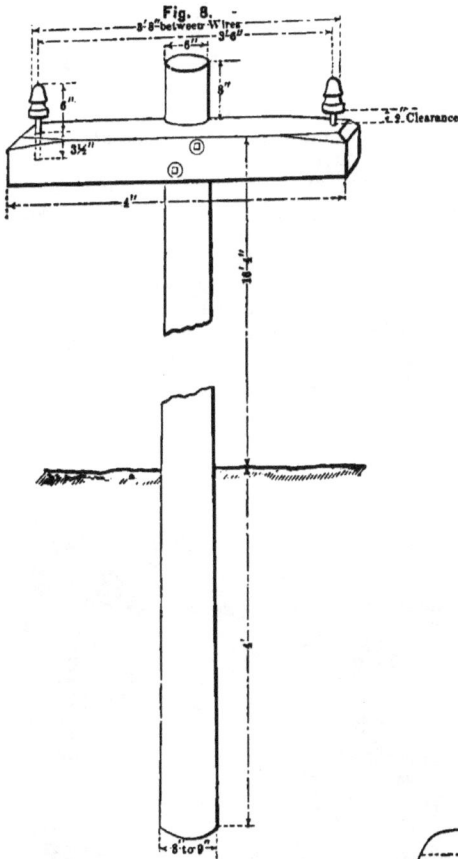

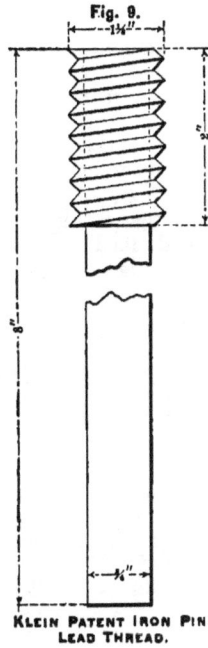

KLEIN PATENT IRON PIN.
LEAD THREAD.

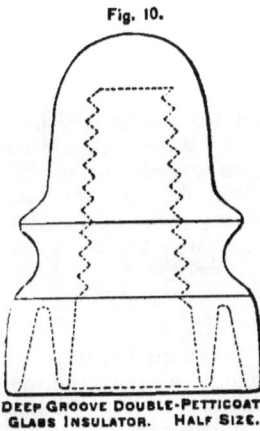

Fig. 10.

Fig. 11.

DEEP GROOVE DOUBLE-PETTICOAT
GLASS INSULATOR. HALF SIZE.

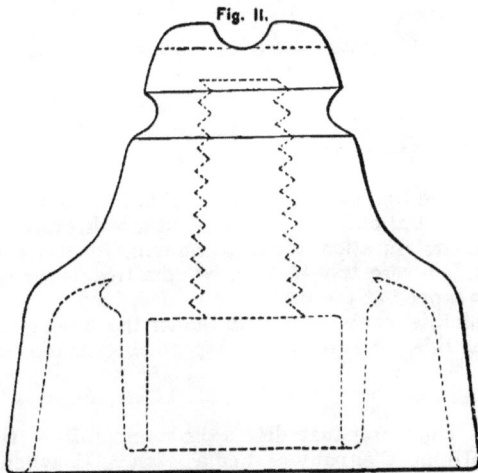

WESTINGHOUSE "POMONA" DOUBLE-PETTICOAT
GLASS INSULATOR. HALF SIZE.

LINE DETAILS.

2*

with an ⅛ in. hole near the end, and in all such places these were used, and held firm by driving a wire nail through them.

The wire was first attached to the insulators by tie-wires of No. 10 galvanized iron wire. Later it was found advisable to insulate the line-wire at the insulators, and for this purpose ordinary sheet-rubber ⅛ in. thick, such as is used for gaskets, was cut into strips 1.5 in. wide and 12 in. long. These were wound spirally about the wire and held in place by two close wrappings of Manson's tape. The whole was then well daubed with asphalt paint, and the insulated wire re-attached to the insulators by tie-wires of No. 6 weather-proof copper wire.

The line crosses a number of very steep ridges (from 300 to 800 ft. in height), and on these the wire necessarily pulls heavily on the top pole, and especially on its pins and insulators. In all such places the ordinary double-petticoat insulators were replaced by the large "Pomona" insulator (Fig. 11), on which the wire is carried in a groove across the top, and its weight is therefore directly down upon and in line with the center of the pin.

Fig. 12.

Pole No. 40; 4,000 ft. from Mill.

(Wire is 17 ft. above ground at pole. Snow-drift 15 ft. deep, March, 1893.)

The line has given no trouble whatsoever, and has carried the high potential of 3,000 volts without a leak, even during a severe storm of ten hours' duration, the rain changing to sleet and ice toward the end; but this severe test, it must be admitted, occurred after the wire had been wrapped at the insulators as described. In fact, one of the chief objects of this insulation was to render the line proof against just such a storm as this. Snow-storms have no effect whatever. (See Fig. 12.)

Motor-Room.

The motor that drives the stamp mill of the Standard Consolidated Mining Company at Bodie is an A. C. synchronous constant-potential machine of 120 horse-power. The mill contains twenty 750 lb. stamps, four wide-belt (6 ft.) Frue vanners, eight continuous-process amalgamating-pans (two of which are constantly grinding), three settlers, one

Plate VI. Motor in operation. Standard Con's Mill

Plate VII. Motor Switch-Board.

agitator, one pan and settler devoted to the amalgamation of concentrates, a bucket elevator, a worm-gear hoist, and a rock-crusher. In order to determine accurately the capacity of motor required, a number of cards were taken with the Tabor indicator from the 20 by 36 inch steam engine that drove the mill, showing an average of 90 and a maximum of $101\frac{1}{2}$ horse-power.

The fields of the motor are self-exciting through a secondary winding on the teeth of the armature, the current being led to a twelve-bar commutator similar to that on the generator. In fact, the motor is almost identical with the generator, the chief difference being in the compensating-winding on the field-bobbins of the latter.

On the armature-shaft of the motor is a friction-wheel, and beyond this a clutch, which is used to set in motion the driving-pulley and the machinery of the mill. On the same bed-plate with the motor is a small 10 horse-power Tesla starting-motor, with a wooden pulley on its shaft, that is brought to bear against the friction-wheel mentioned, by means of a screw and hand-wheel. This Tesla motor consists simply of field-coils and an armature; it has neither brushes, nor commutator, nor sliding contacts of any kind. The alternating current is led directly into the fields, the stationary element, the coils of which, being connected in series, produce a rotating magnetic field, in that each pole is alternately positive and negative. The starting-torque of the armature is, in consequence, very low, and it has to receive several rapid turns by hand before putting on the current, after which it generally runs up to normal speed (1,660 revolutions) within a minute. Plate VI shows the motor in operation.

Turning our attention to the switch-board, shown in Plate VII, the two plugs in the sockets on the right of the board are the line-plugs, and the two to the left of them, in their rests, are the starting-motor plugs. When the line-plugs are in their sockets the current is led directly to the top of the upper jaw-switch, and this switch is never closed until the machines are in synchronism. The wires from the bottom of this switch lead directly to the collector-rings on the armature-shaft of the motor.

In the upper right-hand of the board is the Wurts lightning-arrester, consisting of 22 spools, 11 on a side, separated each by a distance of $\frac{1}{32}$ in. Both legs of the wire-line are attached to the arrester, one on each side at the top, while the ground-wire leads from the bottom spools to a water-pipe in the earth. The spools are made of a patent non-arcing metal, and the dynamo current will therefore not follow the path through them made by a discharge of high-tension atmospheric electricity. The properties of this alloy are such that oxides of the metals are generated by the passage of lightning and not vapor of the metal itself.

To the left of this instrument are two converters of the ratio of 30 to 1, filled with paraffine oil. The primary coil of the right-hand one is connected to the main line just above the plug-sockets, and that of the left-hand converter is connected to the motor-circuit, i. e., the wires leading from the collector-rings on the armature-shaft to the bottom of the upper jaw-switch; it being understood that the motor acts as a generator when being driven by the starting-motor.

The secondary of the line-current converter goes to the top posts, marked G, of the synchronizer (the marble plate with four lamps on it to the extreme left of the board), one leg being first carried through the right-hand side of the lower jaw-switch. When the line-plugs are in, therefore, and this switch is closed, the top light of the synchronizer will always be burning while the generator at the power-house is in operation.

The secondary of the motor-current converter is carried directly to the bottom posts, marked M, of the synchronizer, the two middle lamps of which are connected in series with the motor and generator currents.

. The field-circuit of the motor is carried to the switch-board, and in it are placed the large rheostat, the left-hand ammeter, and the left-hand side of the lower jaw-switch. The closing of this switch and an adjustment of the rheostat will therefore cause the lower light on the synchronizer to burn whenever the motor is being run as a generator, as is the case when it is being driven by the starting-motor. The aluminum fuses showing below the converters are in the main line before it reaches the jaw-switch, as is also the ammeter just below them, which instrument should, and does, record the same volume of current as its fellow in the power-house.

To start the motor requires two men, one to handle the starting-motor and the other at the switch-board. The line-plugs are put in, which leads the main current to the top of the synchronizing-switch, and the lower jaw-switch is thrown in, which closes the field-circuit of the motor, and the secondary of the main-line or generator converter, thereby lighting the upper lamp of the synchronizer. The armature of the starting-motor is turned a few times by hand, and the two left-hand plugs are then pushed into their sockets, leading the current from the main line to the fields of this motor.

Immediately upon doing this, the main-current ammeter records 30 amperes, and the needle stays at this reading until the starting-motor is up to speed, when it drops quickly to 18 to 20 amperes. It takes from fifty to seventy seconds for the starting-motor to reach full speed, after which its friction-wheel is brought to bear against that of the main motor, and the armature of the latter begins to revolve. During this time the synchronizing-switch (the upper jaw-switch) is open, and all the resistance-coils of the rheostat are left in the field-circuit, in order that the armature may more easily be brought up to speed, by preventing the flow of current in the same.

As soon as the armature is above speed, about two thirds of the rheostat is thrown out, permitting 40 or 50 amperes of current to flow, and the lower lamp of the synchronizer to burn. The pushing onto its button of the little switch on the bottom of the synchronizer now connects the two central lamps in series with the motor- and the generator-currents, and they begin to flash in accordance with the phases, and therefore the speeds, of the two machines. As the speed of the motor approaches that of the generator, the wave-phases come nearer coincidence, and these lamps brighten and darken almost simultaneously.

The attendant stands with one hand on the rheostat and the other on the open jaw-switch, watching these waves of light intently, and just as the two lamps darken in unison, he throws in the switch and pulls one of the starting-motor plugs. The lamps only remain "out" for a second or less, while the speeds are together, and then flash up brightly again as the motor speed drops off; there is therefore but a fraction of a second during which the jaw-switch should be closed, though this time can be lengthened slightly by a proper handling of the starting-motor.

If this switch has been thrown in at the right moment, the series lights remain "out," while the top and bottom, or "pilot" lights, burn brightly, and so continue all the while the machines are in operation.

If the switch is thrown in a second or so too soon, the main-current ammeter will fly up to 40 or 45 amperes, and quickly drop down to less than 10 as the motor speed decreases, and it falls into step with the generator, while the series-lamps will remain dark, and the pilot-lamps burn as usual. On the other hand, if it is closed several seconds too soon, or a fraction of a second too late, it is impossible for the machines to get into synchronism. In such event all the lights on the synchronizer go out at once, and a heavy flow of current sets in, the main ammeter showing 45 amperes, which is as high as it can record. By the extinction of the lights the attendant sees at once that he has missed the synchronizing-point, and immediately pulls the main-line plug, opens the jaw-switch, and starts over again. The second trial, however, will not consume as much time as the first, since the starting-motor is still revolving at a high speed, and more quickly comes up to its normal rate, while the motor-armature is also running yet at several hundred revolutions per minute.

This very rarely happens, especially since the addition in March last of an acoustic synchronizer to the phase-lamp device. This instrument emits a sound, the pulsations of which are very rapid at first, the interval between them growing longer as the machines approach equal speed, and settling into a steady hum at the moment of synchronism.

It will be noticed that in order to break the circuit and stop the motor it is necessary to pull the line-plug, on doing which a brilliant arc, sometimes 2 ft. in length, if 25 amperes are flowing, follows out from the socket to the plug-tip. Any attempt to open the jaw-switch while the line-plugs are in would doubtless result in the death of whomsoever tried it, since the distance is too short in which to break the arc, and the current would likely follow down the arm in spite of one's standing on an insulated floor. These floors are used around both generator and motor, and in front of both switch-boards.

The entire operation of starting up the motor from a state of rest occupies from three to five minutes, and when once in synchronism, the clutch can be thrown in and the mill shafting brought to normal speed in from one to two minutes more, after which the load may be thrown on as fast as desired without the least danger of pulling the motor out of synchronism. The clutch is always thrown in slowly in order to prevent too heavy a flow of current, and consequent sparking of the commutator brushes.

By means of a single counter-shaft, fitted with self-oiling boxes, the high speed of the motor (860 revolutions) is reduced to the necessary 80 revolutions of the battery line-shaft, the reductions being 2 ft. to 8 ft., and 3 ft. to 8 ft. Light steel-rim balanced pulleys are used, and an endless 16 in. double leather belt runs from the motor to the first 8 ft. pulley. The speed of this belt is 5,400 ft. per minute, and it is kept tight by levers which, acting through screws, move the entire motor and its bed-plate along four grooved, cast-iron slides.

The motor is separated from the underlying brick foundation by means of 8 by 10 in. timbers, which are bolted to the latter and covered by three layers of 1 in. boards; and to this wood insulation the slides referred to are fastened by lag-screws that pass through the boards into the timbers. The generator is insulated from the I-beams that carry both it and the waterwheels, by timbers 5½ in. thick, to which it is likewise secured by lag-screws.

The mill and offices of the company are lit by 100-volt incandescent lamps, taking current from a large 100-light converter, ratio 30 to 1, which is attached to the main line in the motor-room, before it reaches the switch-board. The light is very satisfactory, even to read or write by, although at times the lamps flicker slightly, due to small changes of load. This variation of intensity is, of course, unavoidable where the lighting current is taken direct from a power circuit; but, in the present case at least, is not sufficiently noticeable to cause inconvenience.

During normal operation of the plant, the field-ammeter of the motor is, by means of the rheostat, kept steadily at 52 amperes, while with full load on the mill the main-current ammeter registers from 23 to 25 amperes. The needle oscillates over a range of 4 to 6 amperes, showing considerable variation of load, due undoubtedly to slipping of the belts, unequal resistance of the grinding-pans, rock-crusher, etc., so that it is difficult to read this ammeter closely, either at the generator or at the motor.

The average amperes of current can be very closely approximated, however, as was done in the tabulated readings given below, which were taken with the object of determining the line-loss, and with the aid of Mr. H. M. Reed, the engineer of the Westinghouse Electric and Manufacturing Company, who installed and first operated this apparatus.

The readings were taken simultaneously at power-house and motor-room, by means of the telephone, and the figures given are the averages of five or six consecutive observations. There being no voltmeter on the motor switch-board, a Weston portable voltmeter was used, the wires being attached to the lower posts on the synchronizer.

The table is merely a rough approximation of the efficiency of the transmission, there being no instruments at hand for close work, such as the measurement of the wheel-shaft energy, or of that given out at the motor-pulley. The efficiency of these machines, namely, of the generator, 95.5 per cent, and of the motor, 93.9 per cent, was determined by Mr. Fred. A. Davis, electrical engineer of the Westinghouse Electric and Manufacturing Company, in charge of the plant. It will be noticed from the table that the line-loss is very light, and also that as the dynamo and motor approach their rated capacity, the efficiency of the transmission increases.

Approximate Efficiency of the Transmission at the Standard Consolidated Mine, Bodie, California.

Date.	Generator. Volts Voltmeter Reading..	Generator. Volts Equivalent to Volts...	Generator. Amperes	Mechanical Horse-Power Given Out by Generator	Horse-Power Given Out by Water-Wheels, allowing 95½ per cent Efficiency for Generator, and adding 5 Horse-Power for Exciter	Motor. Volts Voltmeter Reading..	Motor. Volts Equivalent to Volts ..	Motor. Amperes	Mechanical Horse-Power Delivered to Motor	Line-Loss—Per cent	Mechanical Horse-Power Given Out by Motor, allowing 93.9 per cent Efficiency	Approximate Per Cent of Full Mill-Load Driven at Time of Test	Approximate Per Cent of Water-Power Obtained at Motor-Pulley
December 22d	100	3,390	20	90.9	100.2	103	3,090	20	82.8	8.9	77.7	82	77.2
December 29th	100	3,390	21	95.4	105.0	103½	3,100	21	87.3	8.6	82.0	85	78.2
January 16th	100	3,390	23	104.5	114.4	104	3,120	23	96.2	8.0	90.3	95	79.0
February 1st	101	3,420	25	114.6	125.0	105	3,150	25	105.5	8.0	99.0	105	79.2

NOTE.—During the above test the exciter was being driven from the wheel-shaft instead of by its separate water motor; hence, the allowance of 5 horse-power.

The proper setting of the brushes on the commutator of the motor-armature is a knack acquired only by experience; and for awhile considerable trouble was caused by undue sparking at these brushes. Experience on the part of the attendants has entirely overcome this; but it has been found necessary to use two commutators, keeping one always turned and polished ready for use, and changing them usually after twenty-five to thirty days of steady operation.

In order to stop the motor, the load is thrown off by means of the clutch, and the line-plug is then pulled. Should the plug be pulled without first throwing off the load, a momentary rise in electro-motive force may follow, sufficient to damage an armature-coil. This has happened once in our experience, and the very high potential was vividly shown by the discharge through the lightning-arresters at both ends of the line.

The dependence of the motor speed upon the alternations of the generator is very prettily shown, when, without pulling the line-plugs, the machines are stopped by shutting off the water on the wheels. The motor then slows down in exact accordance with the generator, and is at rest within half a minute or so; whereas, when the plug is pulled in the usual way, the motor-armature will revolve for several minutes from its own momentum before coming to a stop.

This plant has accomplished several unbroken runs, day and night, one of thirteen days' and another of twenty days' duration, but latterly it has been operated more intermittently, on account of the mill being run upon only half-time. During December, 1893, the plant was started twenty-three times in twenty-one days, and in January eleven times in as many days (in accordance with the requirements of the milling work), these daily startings being an excellent test on both the starting-motor and machines, as at such times the differences in potential, and consequently the strain on the insulation, are likely to be a maximum.

The only trouble now experienced with the plant comes from an extraneous source, common, in a greater or less degree, to all electrical plants throughout the world, namely, occasional incursions of lightning during thunder-storms; or from another, more local cause, already alluded to, namely, discharges of static electricity, due to a gradual charging of the line from a highly charged atmosphere during windstorms. These have several times caused the burning-out of armature-coils; but this matter is not as serious as it may seem, since but a couple of hours are required to repair such damage. To put in a new coil, the top-half of the field of the machine, weighing about two tons, is swung off by means of differential blocks and an overhead trolley; the burnt coil is cut out with a hack-saw; the new coil is slipped over the tooth and squeezed into place by means of specially-made geared clamps; the connections are soldered, taped, and painted; and the top-field is then replaced in position.

The entire cost of this plant does not exceed $38,000, while its operation during the month of October alone effected a saving of $2,100, equivalent to $1 46 per ton of ore crushed, and reducing the total milling cost to $2 32 per ton; a fairly low figure for a high-priced camp (wages $4 per day) such as Bodie.

At present the plant is operating most smoothly, and is successfully demonstrating the effectiveness and simplicity of the single-phase synchronous system for such work and distances, while the daily saving over the use of steam, on twelve-hour runs, is from $35 to $40. The writer takes pleasure in acknowledging the aid of his assistant, Mr. R. C. Turner, E.M., in the preparation of the accompanying drawings.

EXTENSION OF THE SYSTEM.

The availability of electric power; the readiness with which, after it is once introduced, it can be applied to any extensions of surface plant or new works such as continually arise in progressive mining, has been recently demonstrated in our experience at Bodie.

For the rapid handling of the tailings, which were to be treated in large vats by a leaching process, it became necessary to put up an incline track about 1,400 ft. away from the mill. To haul the cars up this grade by electric power all that was required was to put in a 15 horse-power direct-current generator at the mill, belt it to the motor, and take current at 500 volts to a 10 horse-power motor at the top of the incline. This is being done, the balance of 5 horse-power to be used in lighting the leaching-plant and operating the necessary pumps.

With electricity this transmission of power to the new plant was extremely simple. By rope or other means it would have been expensive, costly to maintain, and much less efficient.

The following description of the transmission plants of the San Miguel Consolidated Gold Mining Company, of Telluride, Colo., and of the Willamette Falls Electric Company of Portland, Oregon, is taken from "Long-Distance Transmission for Lighting and Power," by Charles F. Scott, being a paper read at the general meeting of the American Institute of Electrical Engineers, Chicago, Ill., June 7, 1892:

"The test of practical operation in long-distance transmission has been applied in but few cases, and the severe test of continued operation over a considerable length of time is of rare occurrence. The latter is the crucial test, and its commercial significance gives it the highest importance. The continued and successful operation of one plant for a year, under extreme conditions of situation and service, is of higher value in testimony to the practical development and possibilities of electrical work than many elaborate projects, or the operation of novel apparatus for a short time.

"It is with this idea in mind that a description is here to be given of two plants, one for lighting and the other for transmission of power. The conditions which have been met include a very considerable distance, extreme difficulties of climate and roughness of country, exacting requirements in continuity of service, and a pressure above that ordinarily used in the class of machines employed. The plants to be described are the first of their type installed in this country, and the apparatus in the power-plant is of a kind that has not been heretofore used. The type and construction of the machines, and the arrangement of apparatus, are new in many particulars, and as they have contributed largely to successful operation they will be described with some minuteness. Alternating-current machinery is employed, constructed by the West-

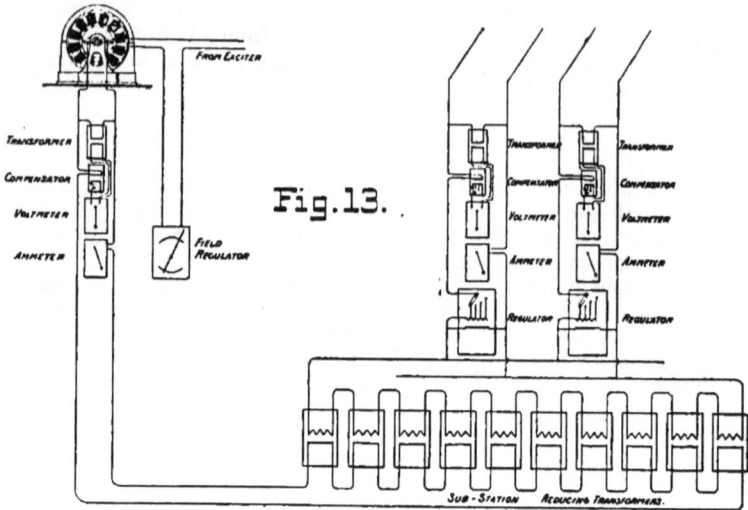

Fig.13.

Lighting-Plant at Portland. Diagram of Apparatus and Connections.

inghouse Electric and Manufacturing Company, the pioneer company in alternating work in this country.

"The lighting-plant was first installed. It is operated by the Willamette Falls Electric Company, of Portland, Oregon. The general requirements are those which electrical transmission is admirably adapted to meet. The falls of the Willamette River, at Oregon City, in the combined points of size, accessibility, and nearness to the seaport, are unequaled. These falls, estimated at from 200,000 to 250,000 horse-power, are about 13 miles from Portland, and it requires but a moment's thought to appreciate the value of an agent which can make this power available in the city.

"The Willamette River is about one quarter of a mile wide, and the fall is about 40 ft. The present station is located on an island at the middle of the river. Victor wheels of 300 horse-power are geared to horizontal shafts, from which the dynamo belts pass to an upper floor at an angle of 45°. Two alternating-current dynamos for incandescent lighting are driven by each wheel. The current, at a pressure of 4,000 volts, passes directly to the line of No. 4 B. & S. wire, which is carried on ordinary double-petticoat glass insulators across the level country to a sub-station in Portland. The current is received at 3,300 volts by transformers in the sub-station and is reduced to 1,100 volts, for distribution by various circuits through the city to ordinary transformers, by which it is reduced to 50 or 100 volts.

"When the apparatus was designed it was not considered practicable to generate 4,000 volts with the ordinary type of machine, in which the wire is wound upon the surface of the armature on account of the difficulty of insulating for more than 1,000 or 2,000 volts. The work was undertaken with a new type of armature, which is specially noteworthy, as it has rendered high potentials practicable in a machine of simple construction.

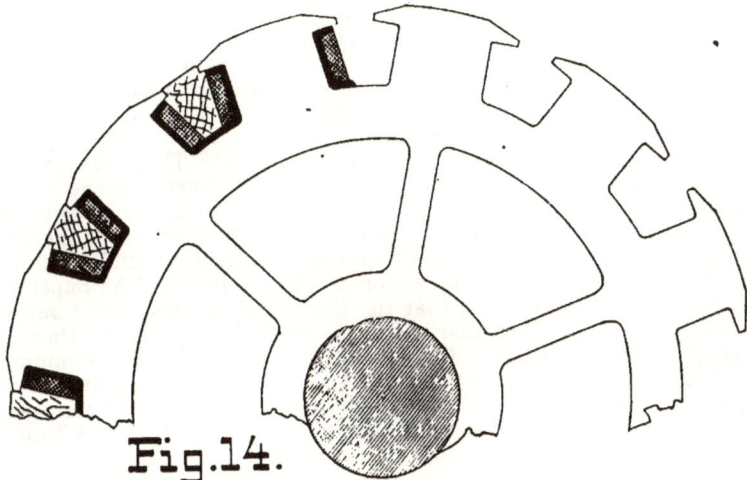

Section of Armature, showing Iron Disk with three Coils in place.

" The field of the dynamo is of the ordinary type of alternating-current machine in use in this country. The casting is circular in form, with twelve inwardly projecting poles of laminated iron, on which the field-coils are placed. This type of machine combines simplicity with rigidity and strength, as both bearings and the lower field are in one casting.

" The armature is built up of laminated disks, which are punched with twelve T-shaped teeth. The armature-coils are wound in a lathe, are carefully taped and insulated, and are then placed over the teeth and sprung in under the projections. The space between adjacent coils is filled by a block of wood, which holds them in place securely. This form of construction gives all the advantages of machine winding over hand work, allows ample insulation between coils and core, protects the coils from mechanical injury, holds them in position without the use of band wires, and makes the replacing of a damaged coil comparatively simple. The field current is supplied from a direct-current machine, and the main current is taken from two collecting-rings on the armature-shaft. The reducing-transformers are placed in a vault in the sub-station at the city. They are arranged in banks or units of ten. Each bank is supplied by a separate dynamo, and has a capacity of 1,250 16-candle-power lights. The coils of the transformers are separately wound and taped, and are separated from one another and from the irons by strips of wood. The primaries are connected in series for receiving 3,300 volts, and the secondaries are in series for delivering 1,100 volts, so that there are 330 volts in the primary and 110 volts in the secondary of each converter. This method of connection throws small differences of potential in any single coil, and permits the use of conductors of good size. The necessity for special insulation is between the coils, where there is ample room for placing it. A transformer may be readily cut out of circuit by short-circuiting its terminals, and in case of an accident in which a coil becomes short-circuited, the E. M. F.

on that transformer disappears, and the others are called upon to do a larger share of the total work without interfering with service. The efficiency of the transformer at full load is 96 per cent.

"The plant was first installed with two incandescent machines, and started nearly two years ago. Since that time five additional machines have been added, so that there are now seven, each with a capacity for supplying 1,250 16-candle-power lights, in Portland. The total capacity is 8,750 lights. The dynamos run admirably. There was one night when several armature-coils were burned out, which was attributed to an iron wire falling across the main line and connecting several of the circuits, grounding them. Otherwise there have been no difficulties to speak of with regard to the operation of the machines. The Superintendent of the plant states that the line has given very little trouble, much less than would ordinarily be expected from a city line. He also says that 'the converters in the sub-station have not given one minute of trouble, and have not cost one cent for repairs.' One explanation of the success of the plant is the intelligent policy of the General Manager, in harmony with his statement that 'It is not the first cost which counts, but the cost of throwing out and replacing apparatus.'

"The same policy has happily governed the installation of the second plant to be described—the power-plant. This is located near Telluride, Colo., and is owned by Mr. L. L. Nunn. The Gold King mill requires power for operating its crushers and stamps, and fuel can come only from long distances at enormous costs. A few miles from the mill there is a water-power, but the country between the two points is steep and rough, and for many months in the year is covered with snow. Electricity is the one means of getting the power from its source to the mill. The conditions are of the most favorable character for demonstrating the value and possibility of electrical transmission.

"In this plant a Pelton wheel, receiving water through a 2 ft. steel pipe, under a head of 320 ft., drives an alternating-current generator. The current is carried over a line of bare wire to the mill, which is nearly 3 miles distant, and drives an alternating-current synchronous motor of 100 horse-power. The generator and motor are machines of the same size and form of construction as the dynamos at Portland, already described, and differ from these only in some minor modifications.

"The generator is provided with a composite field winding. A part of the magnets are excited by direct current from a separate machine, and the rest are excited by a current from the generator armature, which is proportional to the main current, and is commutated by the equivalent of a two-part commutator. The adjustment is such that the E. M. F. on the main terminals rises as the current delivered by the machine increases, compensating for line-losses and keeping the pressure at the motor 3,000 volts. The speed is 833 revolutions, giving 10,000 alternations per minute. The switch-board and regulating appliances are similar to those in the station at Portland. Ordinarily no adjustment is required after the machine is started, and the attendant has little to do besides looking after the mechanical running of the apparatus.

"When the motor is running, the only things to be cared for are the brushes and the bearings. The high-tension brushes—the only point besides the switches where the high tension is exposed—will run for a

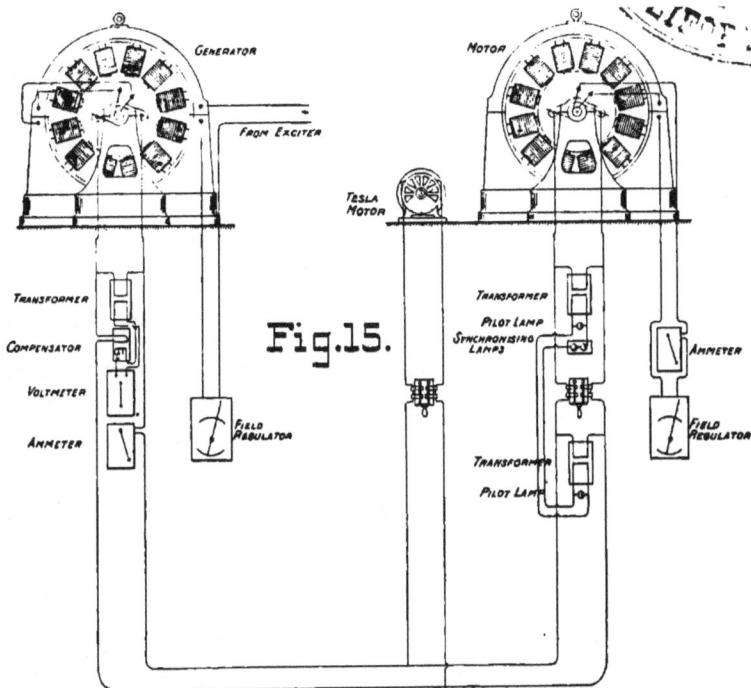

Synchronous Motor-Plant at Telluride. Diagram showing Apparatus and Connections.

week without adjustment, the exciter brushes run without sparking, and the lubrication of the bearings is well provided for. The construction and operation of the motor are strikingly simple in comparison with the steam engine, which it replaces, with its many moving parts and intricate motions.

"A few points illustrating the characteristics of the synchronous motor may be mentioned, as they are of both theoretical and practical interest. The connection of the motor to the generator is not a delicate operation. If the motor is running above synchronous speed at the time of connection with the generator, it instantly adapts itself to the proper speed. If the motor speed is slightly lower than that of the generator, it may fall into step when the switch is closed, but if it be running considerably slower it will not come into synchronism, but will further decrease in speed. When this occurs, the switch of the large motor is opened and that of the starting-motor is closed, bringing the machine up to full speed again without any injury to the apparatus. If the E. M. F. upon each of the machines before connecting them be 3,000 volts it will remain unchanged when they are connected. If the field-current of either machine be increased, the E. M. F. will be raised, but the field-current of the other machine may be lowered and the resulting E. M. F. made equal to 3,000 volts. The current flowing between the machines depends upon the relative field-charges, and is least, whatever the load may be, when the two machines are equally or very near equally excited. The field-current of either machine may be made zero, and the motor will still run, but with greatly reduced capacity. In a test with machines of a smaller size the E. M. F. was 2,000 volts when the two field-charges were equal. When the field-charge of either machine was cut out it fell to 1,200 volts and the current increased very considerably. ·

" The efficiency of the synchronous motor system, leaving out loss in conductors, but including losses in generator and motor in the plant for the delivery of 50 horse-power, was found to be 83½ per cent at full load, and 74 per cent at half load.

" Full load may be thrown on the motor suddenly. In the Gold King mill the stamps, which are operated by the motor, are usually left raised when the plant is stopped, in order to avoid the extra strain of lifting them when the plant is started. It sometimes happens that the stamps are left down and the motor is required to raise them all at once. When the clutch is thrown in, the current indicates that the load is considerably above the normal capacity of the motor, and yet it is started without difficulty or apparent strain.

" The excellent current regulation with different loads, the tendency of the machines to normal adjustment when there is ordinary variation in the field-currents, the small liability to injury when the motor is greatly overloaded, the high efficiency and ease of attendance, are points of great value in the practical operation of the system.

" The pole-line runs from the power station up the mountain to a height of 2,500 ft., and then crosses a rough but comparatively level country to the mill. The line at some places is at an angle of 45°, and many of the poles had to be set in solid rock. The surface of the snow in winter is occasionally at a level with the tops of the poles, and parts of the pole-line are practically inaccessible during some months of the year. This region is peculiarly subjected to lightning discharges, and special precautions are necessary to protect the apparatus. In one instance there were forty-two discharges of the lightning-arresters in as many minutes.

" The plant was started for regular work in June of last year. An accurate record was kept from the middle of July to the first of May, showing the actual number and the length of the delays caused by electrical machinery. During these nine and a half months the system was in regular continuous operation six and a half days each week, with but few intermissions. The difficulties which were encountered were insignificant in amount and have resulted, not from any fundamental difficulty in the system, but have been caused by incidental defects or accidents which usually indicated their own remedy. The stops due to the electrical machinery resulted from a variety of causes, and comprised the replacing of an armature-coil damaged by lightning, renewing of fuses, fixing loose contacts, the examination of the line after a storm, and sundry other slight mishaps. The aggregate time lost on account of the electrical apparatus was, by actual count, less than 48 hours during three fourths of a year. A recent report from the Superintendent of the plant, covering the time from December 13th to May 1st, shows that the plant was running 127 days with a loss of 19¾ hours, or, as he puts it, an average of about nine minutes in a day of 24 hours. Although the plant was generally shut down each week for 12 hours on Sunday, this was not practicable during a part of the winter, and the motor on

Plate VIII. Machine used as Generator and Motor at Telluride, Colorado.

Plate IX. New Motor at Telluride, Colorado. 250 horse-power.

one occasion was run continuously for 27 days without any stop whatever. Such a record as this, with a new type of machinery, in a country, where line construction and maintenance are peculiarly difficult, with practically continuous service, with attendants who were not electricians, with a high voltage, a considerable distance and large power, places transmission by the alternating-current synchronous system beyond the stage of experimental trial and gives it the stamp of commercial success.

"This success is confirmed in a substantial way by the immediate extension of the plant. A 50 horse-power motor is now being installed at a mill a few miles from the Gold King; an order has been entered for a 750 horse-power generator to be located in the power station; and a 250 horse-power motor for operating a mill about 10 miles distant. Lighting at Telluride, 8 miles from the station, which has been done heretofore on a small scale on a circuit from the power generator, is being extended.*

"The large generator is a new design and is notable, as it has more than three times the capacity of any alternating-current dynamo previously made in this country. Two machines of this size have been running for some months for incandescent lighting at St. Louis.

"This dynamo is of a type similar to the machines at Portland and Telluride. The field has twenty-eight poles, requiring a speed of 570 revolutions for 16,000 alternations, the conditions of running at St. Louis, and 357 revolutions for 10,000, as it will be operated at Telluride. The armature has T-shaped teeth, as in smaller machines. The diameter of the armature is slightly over 4 ft. and its length is about 2 ft. There is a third bearing at the end of the shaft outside of the pulley to relieve the other bearings from the severe strains resulting from belt tension. The total height of the machine is 8 ft., and its weight is 40,000 lbs. The electrical efficiency at full load is over 95 per cent.

"The extension of alternating-current working, both for lighting and power, by the use of large machines, is therefore already provided for.

"The extension to greater distances is largely a question of E. M. F. Nearly every one is familiar with the rapidity with which the cost of copper diminishes as the voltage is increased. If the cost be $100 with 500 volts, it will be $25 at 1,000 volts, and $1 at 5,000 volts. The higher the tension, however, the greater the difficulty and cost of construction, and the greater the liability to accident with apparatus and line. There are a few points in connection with this subject which may be noted without entering into a general consideration of it.

"The smallest size of wire that can well be used for line work on account of its mechanical strength is about No. 6 B & S. This wire will transmit with 20 per cent loss 100 horse-power 10 miles at 4,000 volts, or twice the power half the distance. Unless these distances or powers are to be exceeded, an increase in pressure would result in no saving in copper, but simply in a less line-loss, which is already not excessive.

* NOTE, May, 1893.—Since this paper was read the 750 horse-power generator, the 50 horse-power motor, and the 250 horse-power motor have been installed and put in operation. (See Plates VIII and IX.) The order has been placed for an additional 75 horse-power motor. A letter from the engineer of the plant, dated April 20, 1893, contains the following: "The generator runs very nicely indeed. The motors are running very satisfactorily. The 250 horse-power is running very nicely at the present time. The 50 horse-power has been running about a week since the winter shut-down, and the 100 horse-power motor, which has been idle during repairs to the mill, is expected to start within two weeks. When we get the 75 horse-power motor and start it, we will feel that we have quite a system in operation."

"The use of 4,000 volts at the motor and a line-loss of 20 per cent requires an outlay for copper of only about 10 to 15 per cent of the total cost of the plant when the distance is 10 miles. Unless, therefore, the cost of copper is to bear an insignificant proportion of the total cost, it is unnecessary to exceed this pressure unless the distance be greater than about 10 miles.

"These simple considerations show that pressures practically the same as those employed in the plants which have been described are ample for considerable distances. The same type of apparatus which has been successful in them is available for larger capacities. The fundamental elements required for electrical transmission in a very wide range of cases have therefore been tried and their success demonstrated.

"For considerably longer distances, where pressures higher than about 5,000 volts are required, good practice indicates the use of transformers for raising the pressure at the generator and reducing it at the motor, similar in general to those employed at Portland. The increased pressure thus available greatly reduces the cost of copper required, and this reduction must, of course, be more than sufficient to cover the cost of the transformers.

"The simplicity and flexibility and range of the alternating-current system make its possibilities the sole dependence of the largest enterprises toward which the public and engineers are looking. The records of the plants at both Portland and Telluride demonstrate that these possibilities are being realized, and that work in this field is fast passing from experimental investigation into practical electrical engineering."

PROTECTION AGAINST LIGHTNING.

The protection of these circuits against lightning is a very important matter. The mountainous regions in which the majority of mines are located are generally subject to severe thunder-storms and electrical disturbances, and therefore to the miner using electrical power, it is of vital importance to have his apparatus protected from damage from such causes; while many may be deterred from undertaking the installation of electric transmissions through fear of lightning interfering too seriously with their successful operation.

During the past year much trouble has arisen at Bodie from this cause; the machines running smoothly and perfectly for several weeks, when a storm would occur, oftentimes without visible lightning or audible thunder, but causing a burn-out of armature-coils in generator or motor.

Lightning-arrester houses have been built at each end of the line close to the power-house and motor-room, and these fitted up with banks of spark-gap-arresters and choke-coils precisely as outlined by Mr. Alex. J. Wurts in his article on " Discriminating Lightning-Arresters and Recent Progress in Means for Protection against Lightning " (in the " Electrical Engineering Magazine," May 23, 1894), since which time but little trouble has been had.

Our experience was so similar to that of Mr. Wurts with the plant of the San Miguel Consolidated Gold Mining Company, at Telluride, Colo., and so much interesting and valuable information on the subject is given in the above-mentioned article, that the writer takes the liberty of quoting Mr. Wurts in full:

An Experiment with Lightning-Arresters on a 3,000-Volt Alternating-Current Circuit.*

"During the winter of 1892 and 1893 I made a searching investigation of this subject, experimenting with disruptive discharges and various kinds of combinations of apparatus which might promise advantageous results, and since that time have spent nearly six months in the State of Colorado—a land of thunder-storms—testing the various forms of apparatus which I had designed as a possible protection against lightning.

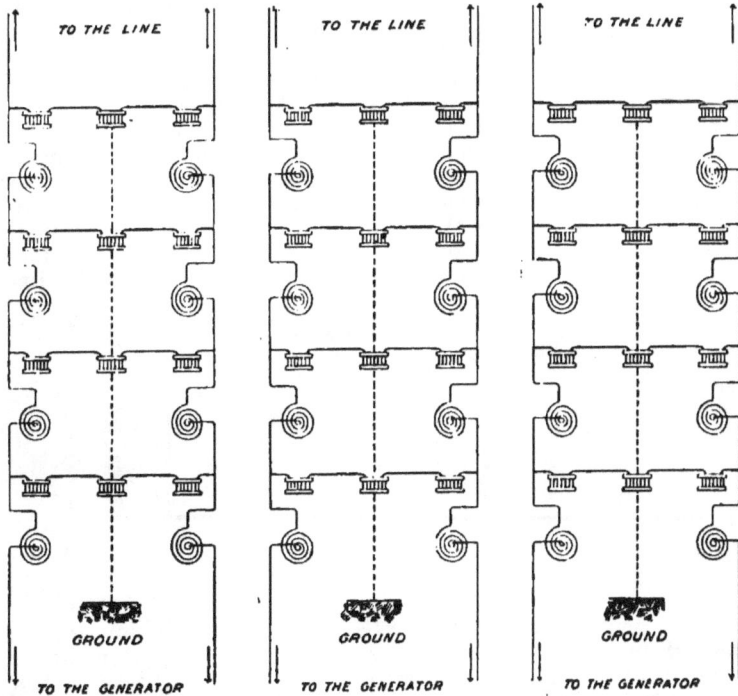

Fig. 16.

"The general requirements of efficient lightning-arrester apparatus are: (1) To provide discharge circuits which shall operate automatically and repeatedly, and which shall with certainty avoid dynamo short-circuits or interruption of the system. (2) To provide discharge circuits, or so install them that they shall invariably offer a certain path to ground for disruptive discharges in preference to any other part of the system. It follows also from this last, and as a matter of practical experience, that ground-discharge circuits should be short and straight,

* Abstract of a paper read at the eleventh general meeting of the American Institute of Electrical Engineers, Philadelphia, May 15, 1894.

3*

and that ground connections should be of the most approved construction. Experiments were therefore made to determine the number of non-arcing metal cylinders and spark-gaps which would be necessary to interrupt a short-circuit on a 3,000-volt alternator with the potential raised to 3,300 volts. Nineteen cylinders, or eighteen gaps, were found to offer ample margin, and the breaking down E. M. F. on half this number of gaps, which would intervene between line and ground, was found to be about 70 per cent of the E. M. F. required to break down insulation ordinarily used in a 3,000-volt generator. The form finally adopted was that of a flat coil about 18 in. in diameter, and wound with seventeen turns of wire; the size of the wire varied, of course, with the carrying capacity of the particular circuit into which it was to be connected. After further experimenting with various combinations of spark-gaps and choke-coils, it was decided that the trial apparatus could consist of eight choke-coils and twelve 1,000-volt non-arcing metal-arresters for each end of each circuit; that is, four choke-coils should be connected in series in each leg of each circuit, with discharge circuits intervening. The relative positions of these parts are clearly indicated in Fig. 16, which represents one end of each of the three circuits. Disruptive discharges form nodal points in the system; that is, points where there will be a minimum tendency to discharge; hence to avoid these with any degree of certainty a multiplicity of arresters, preferably line-

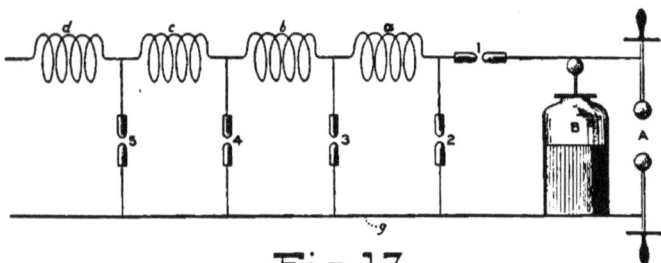

Fig. 17.

arresters, should be used. Choke-coils form points of reflection, or points where there will be a maximum tendency to discharge; hence, a discharge spark-gap connected directly in front of a choke-coil is more likely to receive discharges than if placed at some other point in the system. Disruptive discharges are liable to divide and follow several paths, being governed by a complexity of ever-varying circumstances. For this reason it was decided to connect several choke-coils in series, so that should only a portion of a discharge pass across the first arrester, the balance passing through the first coil, a second opportunity for discharge would be found at the second arrester, which was also connected in front of a coil, and therefore at a point of reflection. Should this remaining portion of the discharge again divide, a further opportunity for discharge would be afforded at the third arrester, and so on, so that by the time the fourth coil was reached it was presumed that the discharge would have spent itself.

"Some of the many experiments made to substantiate this theory are exceedingly interesting as well as instructive. Referring to Fig. 17, A are the terminals of a powerful influence-machine, B is a battery of Leyden jars, G a wire which may represent the ground, and is connected to the outside coating of the jars; L is a second wire which may represent one leg of an electric circuit; a, b, c, and d are choke-coils connected with the line L, and in series with each other; 2, 3, 4, and 5 are intervening discharge-circuits containing spark-gaps; 1 is a ¹⅜ in. spark-gap separating line L from the inside coating of battery B. If now the battery becomes charged from the influence-machine A, a large and violent disruptive discharge will take place across gap 1, and suddenly charge line L. This discharge will then pass to earth G through one or more of the spark-gaps 2, 3, 4, 5, according to circumstances. It will be noted now, that this arrangement of choke-coils and discharge-circuits is similar to that shown in Fig. 16. The spark-gaps were made of rounded ½ in. brass rod adjustable with a ¹⁄₃₂ in. screw thread. The coils were wound with No. 0000 wire, were 3 in. in diameter, 6 in. long, and contained eleven turns each.

"With these four choke-coils in series, and spark-gaps intervening, the discharges are so thoroughly sifted out that only an occasional thread-like spark finds its way across the last gap. With laboratory results such as these, it seems fair to presume that results more or less similar might also be expected in practice.

"The plant selected for the trial of this apparatus was that of the San Miguel Consolidated Gold Mining Company, of Telluride, Colorado, which is equipped with a 3,000-volt alternating-current synchronous system, of 1,000 horse-power capacity, operating stamping mills, and furnishing current to the Telluride Electric Light Company. These points are situated among the mountains at distances varying from 3 to 10 miles from the power-house. Three separate circuits leaving the power-house extend over a wild and rocky country, and in some places rise above timber line. In previous years every attempt to protect this plant from lightning had failed. During the summer months two horses were kept constantly saddled ready for emergencies consequent on lightning discharges, and at the motor and power-house it was common practice, on the approach of a thunder-storm, to lay out, ready for instant use, an extra armature-coil, with all the necessary tools for handling the same. In one of the former types of arresters used in this plant forty fuses were blown inside of sixty minutes.

"The apparatus was then installed in a specially constructed and weather-proof lightning-arrester house, the arresters and choke-coils being mounted on thoroughly dried wooden frames, and every precaution taken to insulate these from the ground and from each other. In this manner, also, it was expected that the lightning discharges would be kept entirely out of the station. The connections inside the power-house lightning-arrester house are all clearly indicated in Fig. 16. One main B. & S. No. 3 ground-wire was used for all the arresters in each bank, unnecessary kinks and bends being studiously avoided.

"Observations were taken by competent men at each bank of arresters during the entire lightning season, and the results obtained indicate that the discharges occurred most frequently over the second arresters; many passed over the third arresters; very few, however, over the first or fourth. The writer personally watched one of these banks of arresters through severe thunder-storms, and in every instance the discharges noticed by him were seen to pass across the second series of spark-gaps, but in no instance was there a fuse blown or damage done to the arresters."

The matter of proper earth connections at the terminals of the ground-wires from these lightning-arresters is most important. The common idea that a rod of iron stuck into the earth a couple of feet, or a piece of old iron thrown into the bed of a stream, forms a good "ground" is most erroneous, as has been proven by our experience at Bodie. In reality, such form the poorest kind of earth connection.

The ground-wires from the banks of arresters are carried into a pit dug directly underneath the arrester-house. This pit is about 4 ft. square and reaches to permanently moist earth at the power station, while at the mill, since no damp ground was found at a reasonable depth, it was necessary to carry a ¼ in. pipe into the pit through which water is occasionally allowed to run. On the bottom of the pit a layer of small charcoal 2 ft. in depth was laid, and on this placed a plate of copper $\frac{1}{16}$ in. in thickness and 6 sq. ft. in area. To this plate the ground-wire was securely soldered in a spiral coil, and another layer of charcoal placed on top of it, after which the pit was filled up with loose dirt. This makes an efficient ground, and gives the necessary surface for the rapid dissemination of the electric discharge.

Inductive resistance-grounds have also been added at either end of the line, with the object of preventing damage to the machines through static charges, by leaking them to earth as fast as accumulated. They are connected to each side of the line as shown in sketch, and their resistance is sufficiently high to prevent any appreciable loss of current.

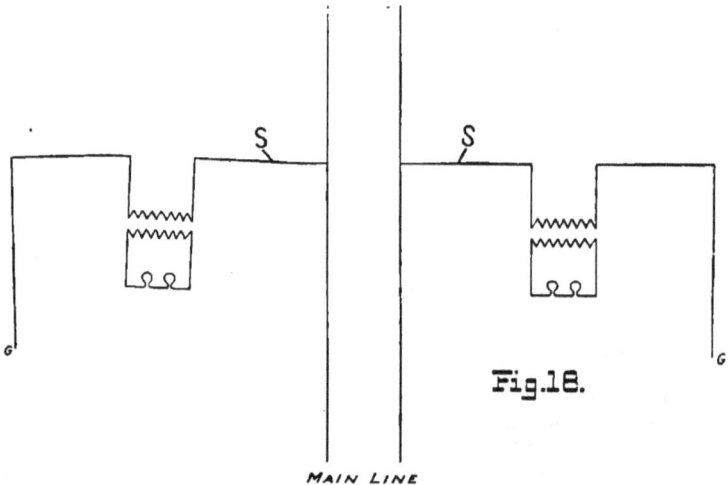

Fig.18.

MAIN LINE

Two 50-volt lamps in the secondary circuit of each "resistance" burn at a dull red when the switch S is closed; as is the case at all times excepting during thunder-storms. During such these grounds are taken off in order to prevent danger of burning out the primaries.

These inductive resistances are chiefly of service during wind-storms, when the line is most liable to cumulative charges from the atmosphere.

ELECTRICAL TRANSMISSION PLANTS AT PRESENT IN OPERATION.

The following short descriptions of electrical power plants in operation have been culled from various sources, many of them from the article by W. H. Adams in "Engineering and Mining Journal," June 23, 1894:

Electric Generating Station at Tivoli.—In a lecture delivered by Prof. J. A. Fleming, at the Royal Institution, the following description was given of a plant used in transmitting 2,000 horse-power from the Falls of the Anio, Tivoli, for 18 miles over the Campagna to Rome: "From the upper levels of the Anio an aqueduct has been led which delivers water to the top of an iron pipe 150 ft. above the power-house. This power-house is placed about halfway down the declivity on which are situated the famous cascades of Tivoli. The pipe is about 2 meters in diameter and can deliver 100 to 150 cu. ft. of water a second with a head of 150 ft., or nearly 2,000 horse-power. The water is conducted to a series of nine Girard turbines, six being of 350 and three of 50 horse-power. The six larger ones are directly connected with Ganz alternators, which generate a current of electricity at a pressure of 6,000 volts, while the three smaller ones are used to drive the exciters. The current is conveyed to Rome by four cables carried on 760 posts, which are placed in a straight line across the Campagna. Outside the Porta Pia, at Rome, it enters a transformer-house, where its pressure is reduced from 5,000 to 2,000 volts. Part is then used for arc lighting in the streets of Rome, and the rest is distributed by underground cables to various other centers, where it is again transformed down to a pressure of 100 volts for use in houses. About 20,000 incandescent lamps are thus supplied with current."

Compagnie de l'Industrie Électrique, Genève, Switzerland.—Four hundred horse-power transmitted 20 miles by continuous currents, potential between 6,000 and 7,000 volts. One 400 horse-power turbine, under fall of 14 meters, revolves on vertical shaft at 120 revolutions per minute, and drives by means of bevel wheels and pinions two dynamos, placed one on each side. Commercial efficiency of the dynamos reaches 93 per cent, with a weight less than 7½ tons. Commercial efficiency of the installation from shaft of the turbine to the motor shafts exceed 75 per cent at full load. Bare copper conductors are used in line construction, 7 millimeters in diameter, the line being entirely aërial through mountainous country between Frinvilliers and Bieberist.

A plant at *Oynuax, France,* has been working satisfactorily for some time with two turbines of 150 horse-power each. A generator of 105,000 watts capacity at 2,000 volts is directly connected to each turbine. Distance between generating and receiving station about 8 kilometers, and 76 per cent efficiency is obtained in the transmission.

At *Chambéry,* 2,000 horse-power is about being installed with waterfall 2,040 ft. high. There are to be seven alternators, each of 120 kilowatts at 5,000 volts.

The two waterfalls about 25 miles from *Christiania, Norway,* are about to be utilized for power transmission, at a total cost of $1,500,000. The voltage will not exceed 20,000, to be carried on bare wire on poles carefully guarded over the entire distance.

The *Portland General Electric* power plant, at Oregon City, Oregon, 12 miles from the city of Portland, on the Willamette River, will install 12,000 horse-power, using twenty Victor 42 in. and twenty Victor 60 in.

turbines, under 30 and 48 ft. head. Many new features in tri-phase transmission are promised for this plant, one half of which is now being installed for service during the coming year.

In the Baltic mill, on the Shetucket River, about 5 miles above *Taftville, Connecticut*, the General Electric Company has lately installed a three-phase power transmission plant of 1,500 horse-power, using three double 42 in. horizontal turbines, developing 800 effective horse-power each at 157 revolutions per minute, and one double 27 in. turbine developing 300 horse-power at 244 revolutions per minute.

The efficiency is just 80 per cent from power applied to dynamo pulley to delivery at motor pulley at *Columbia, South Carolina.* The Columbia Cotton Mills Company is about starting a plant of 1,400 horse-power, using two pairs of 48 in. Victor turbines on a horizontal shaft, and a single 24 in. turbine for fire pump. The 48 in. turbines are connected together, and at each end are directly connected to a generator of 700 horse-power capacity. This is the second instance in this country where an entire cotton mill is driven by electricity. The generators, made by the General Electric Company, weigh about 100,000 lbs. each; the armature is 10 ft. in diameter, 500 kilo-watts capacity, and operates at a speed of 108 revolutions per minute.

Concord Land and Water Power Company, Concord, N. H., has utilized 2,000 horse-power of the 5,000 horse-power, furnished by the new dam located at Sewall's Falls. Horizontal turbines are used with draught tubes, thus avoiding gears, the power being transmitted from each pair of wheels to the shafting in the generator-room by belts. The shafting is arranged with quills and clutches, in order that any wheel or section of shaft may be run independently of any other. The pulleys used on shafting are extra heavy, and fly-wheels are being tried for the first time at this location for inertia regulation. Six tri-phased generators are to be installed, two being now in operation, of 250 kilo-watts capacity and separately excited, and run at a speed of 600 revolutions per minute. The current is generated at 2,500 volts; the line runs to the center of the city, about 3 miles, where tri-phased current of a frequency of 50 is delivered to the mains at 2,200 volts pressure. It is then transformed to 110 volts for delivery to consumers, being sold by meter at 20 cents per kw.-hour for lighting and 10 cents per kw.-hour for power, heating, and cooking.

Colorado.

The Roaring Fork Electric Light and Power Company, Aspen.—Pipeline, 500 ft. of 16 in., 3,500 ft. of 14 in. Power plant: eight 24 in. Pelton wheels, 1,000 revolutions, under a head of 820 ft., equal to 175 horse-power each; total, 1,400 horse-power. The light plant supplies the entire town of Aspen, as well as many mills, mines, and sampling-works. The power plant supplies 120,000 watts, and is used for operating mills, hoists, pumps, and tramways within a radius of 3 to 4 miles from the generating station. The plant has been in continuous operation for five years, with practically no expense in the way of repairs or interruption of the service.

Aspen Mining and Smelting Company's Plants.—Flume, 1,300 ft. long, head 80 ft.; two 50 horse-power Thomson-Houston dynamos, equal to 100 electric horse-power; generating station, 6,000 ft. from tunnel entrance. The underground motors are located 1,000, 1,200, and 1,800 ft. from the entrance. Power used for hoisting.

New works of the *Roaring Fork Company*. Two pipe-lines: 2,500 ft. of 26 in. pipe, head 312 ft.; 4,300 ft. of 24 in. pipe, head 330 ft. Power plant: five 60 in. Pelton wheels, 30 revolutions, 250 horse-power each; total, 1,250 horse-power. Distributes power 5 miles distant.

People's Electric Light and Power Company, on Castle Creek, 1 mile from Aspen. Power plant: two 5 ft. double-nozzle Pelton wheels, 300 horse-power each, 240 revolutions, 180 ft. head; also two 3 ft. double-nozzle Pelton wheels, 75 horse-power each, 345 revolutions. Power and light furnished to mills and mines within a radius of 3 miles.

Virginius Mine Plant.—Pipe-line, 4,000 ft., head 485 ft. Power plant: two Pelton wheels, one 5 ft. and the other 6 ft. in diameter, 500 and 700 horse-power, respectively; total, 1,200 horse-power. Electric generating plant: 293 horse-power; line, 4 miles long. Machinery operated at the mines: 2 pumps (one 60 horse-power and the other 25 horse-power), one 15 horse-power blower, two 60 horse-power motors for running concentrators and stamp mills. Previous to installation of the electric plant the outlay for coal alone was $40,000, at $18 per ton.

Telluride—San Miguel Consolidated Gold Mining Company's Plant.— Power plant: 6 ft. Pelton wheel, 3,900 ft. of 24 in. pipe-line, 320 ft. of head; 1,100 horse-power dynamos, supplying power to three stamp mills, 2, 3, and 10 miles distant, and also lights for the town of Telluride, 8 miles distant. The pole-line is 8,800 to 12,000 ft. above sea-level. The cost of maintenance, including wages at power plant, was given at $3,060 for the first year. The repair account was $21, occasioned by lightning. Since the introduction of lightning-arresters there has been no damage from this cause.

Sheridan & Belmont Company.—Water-head, 235 ft.; Pelton wheel, 28 in.; electric circuit, 12,300 ft., furnishing 250 to 300 lights, and two motors of 10 horse-power and 5 horse-power.

Belmont Consolidated Mining Company.—Head of water, 670 ft., capable of developing 210 horse-power, with a 36 in. Pelton wheel. In the mine are two 30 horse-power motors. Length of line, 2 miles. Loss between generators and motors, 8 per cent.

Washington.

Walla Walla Electric Power Plant.—Pipe-line, 5,800 ft. of 48 in. pipe; two 80 in. Pelton wheels; two A-100 Edison 2,000-volt machines.

Idaho.

Cœur d'Alene Silver and Lead Mining Company.—Pipe-line, 3,000 ft.; head, 850 ft.; two 3 ft. double-nozzle Pelton wheels. Has replaced all steam machinery at the mine, with a saving of $40,000 per year.

Mr. Clark, General Manager of this company, writes in reference to this plant as follows: "In respect to the relative merits of steam and electricity at the Poorman Mine, I will say that the amount saved in fuel is about $100 a day. This, of course, is due to the fact that we generate electricity by water power. How electricity would compare with steam in the matter of cost, if the former was generated by steam power, I am not prepared to say, but am of the opinion that where steam has to be transmitted a long distance underground, particularly where it is wet, that electricity generated with steam and transmitted to the pumps or other machinery will be found to be the most economical, the percentage of loss in transmission being so much less; in addition to this, the cumbersome steam pipes, with their destructive effect on shaft timber, is avoided. We have five machines in use: two 175 K.-W. at the generating station 1½ miles distant from our works, where they are operated with Pelton wheels under 800 ft. head; one 175 K.-W. to drive our concentrator; one 150 K.-W. T.-H. machine for the pump, raising 500 gallons of water per minute 500 ft.; and one 175 K.-W. for the compressor. This system has been almost two years in operation, and my experience in that time is that an electric machine to run continuously, as in operating a mine pump or mill, must have at least double the capacity it would require when stops occur—as on a street car, for example."

The greatest departure, however, electrically, is the installation at this mine of an 80 kilo-watt, *1,200-volt* motor for driving a Knowles double-acting pump, having a capacity of lifting 500 gallons 500 ft. high per minute. The current for the motor is conducted down the shaft through which all the steam pipes, air pipes, etc., are taken, by two Siemens (lead, iron band, iron wire, armored) cables—C. L. A. T. W. The iron wire armoring of these two cables is so connected and arranged as to prevent any shocks due to static charges, should workmen come in contact with the cables.

The regulation of the speed of the motor is effected by placing in armature-circuit of motor a rheostat of two or three ohms capacity. The motor drives the pump through a counter-shaft and wooden-toothed gear and bronzed pinion. The crank-shaft of pump makes from 36 to 46 revolutions per minute as desired.

This is the first installation of a 1,200-volt motor placed in a mine for pumping purposes. The mine and mill are lighted by incandescent lamps from a 110-volt Edison dynamo belted from main shaft of mill.

California.

The Dalmatia Mine Plant in El Dorado County.—The power station is located on Rock Creek, some 1,500 feet below the mine and mill, and 2 miles distant in a straight line. The plant consists of an 8 ft. Pelton wheel, which, running under a head of 110 ft. at 100 revolutions, with a 5½ in. nozzle, has a maximum capacity of 130 horse-power. To this wheel is connected a 100 horse-power constant-current Brush generator—30 amperes—speeded at 900 revolutions, the current from which is carried to the mill through a single insulated copper wire, No. 3, B. & S. gauge, the return being made by a wire of the same size, making a 4-mile circuit. The power from the generator is communicated to the counter-shaft of the mill by a 70 horse-power motor running at 950 revolutions. The machinery operated consists of three Huntington mills, a 10-stamp battery, and a rockbreaker. The Pelton wheel under these conditions shows an efficiency of 86 per cent, while about 75 per cent of the power thus generated is available for duty at the mill. Sufficient power is taken from the main circuit to run sixty incandescent lamps for lighting the works, office, and residence of the manager. The mill handles an average of 4,000 tons of ore a month, effecting a saving of some 60 per cent over the former method of working by steam power, while the cost of maintenance is about as six to one in favor of electricity. An extension of this line has recently been made to the St. Lawrence

Mill, similar in character to the Dalmatia—located 3 miles from the latter and 5 miles from the waterwheel station—which is operated by Keith generators and motors. This is an exceptionally long distance for a continuous-current transmission.

The San Antonio Electric Light and Power Company in Southern California.—The power plant is located in San Antonio Cañon. The water is brought to the power station through 1,900 ft. of 30 in. and 600 ft. of 24 in. double-riveted sheet-iron pipe, giving 300 ft. effective head or running pressure. The power station is provided with four double-nozzle Pelton wheels, 34 in. in diameter, coupled direct to the armature shafts of as many Westinghouse alternating-current generators of 200 horse-power each. The wheels run under above conditions 600 revolutions per minute, giving the same speed to the generators. Two exciters are provided, which are also run by Pelton wheels coupled to the shafts in the same manner, and of 20 horse-power each. The current thus generated is carried on two No. 7 bare copper wires 7 miles down the cañon to a point where they diverge, one running to Pomona, 15 miles, and the other to San Bernardino, 28 miles, covering by the return circuit in the latter case a distance of 56 miles. By means of transformers the potential is raised at the generating station to 10,000 volts, and the current carried at this pressure to sub-stations located just outside the cities named, where, by means of step-down transformers, it is reduced to about 1,000 volts and then distributed for both light and power purposes.

Amador County—The Gover Plant.—A 3 ft. Pelton wheel, 340 ft. head, speed 470 revolutions, works two Dow pumps of 15 horse-power and 20 horse-power, which handle 200,000 gallons of water per day.

Plant at Redlands, San Bernardino County.—This is a three-phase plant, recently installed by the General Electric Company; distance of transmission, 5 miles; two A. C. generators, of 250 K.-W. each, driven by four 30 in. Pelton waterwheels, at a speed of 600 revolutions. The generators carry a potential of 2,450 volts and the motor about 2,150, the line-loss being approximately 12 per cent. The three No. 0, bare copper wires (insulated within city limits) are carried on deep-groove, double-petticoat glass insulators. The line poles are 35 ft. long, 6 ft. in the ground, and set 110 ft. apart. The one motor at present in operation is a synchronous high-potential machine of 150 horse-power, and has continuous work to perform in driving the ice machines of the Union Ice Company. The initial current in the fields of the motor is generated by a small exciter, and the motor is self-starting only under light load, the full load being thrown on after the machine is up to speed.

Nevada.

The Chollar Plant.—This, one of the earliest applications of electricity to mining work, has already been so fully described in the mining and technical papers that it is not necessary to repeat the details.

Arizona.

Plant of the Commercial Mining Company.—This plant consists of a 4 ft. Pelton wheel, which runs, under a 1,200 ft. head, at 699 revolutions a minute, developing 45 horse-power, using a nozzle tip $\frac{53}{100}$ in. in diameter; also a 24 in. Pelton wheel running, under the same head, at 1,380 revolutions, developing 20 horse-power, with a nozzle tip $\frac{35}{100}$ in. diameter. These wheels run a concentrating and smelting plant, including rock-breaker, blowers, pump, etc. The pipe-line is 20,000 ft. in length, the upper end being 6 and 5 in. casing, and the lower end 5 in. lap-weld pipe.

ELECTRICITY IN UNDERGROUND OPERATIONS.

This power was applied earlier and has been more widely used in the coal measures than in lode or precious metal mining; hence, in the former we find many coal mines fitted with extensive and efficient electric haulage systems and lit by electricity, while their drills and coal-cutters are operated by the same power.

Electric locomotives are now built no larger than the cars they are to haul, and made to conform to any gauge of track from 18 in. upwards, effacting a great saving in the size of entries required for narrow coal seems. That they do not vitiate the air as do steam motors is a great point in their favor.

For the operation of undercutting machines, haulage, and general coal mining work, the direct-current is usually employed, since it can be generated cheaply at the pit's mouth, and the underground workings are not usually of such length as to increase unduly the cost of conductors. It is not invariably employed, however, as instanced in the First Pool Mines at Benola, on the Monongahela River, where an A. C. three-phase plant was installed several years ago. The generator is of 100 horse-power, Tesla type, operating under 500 volts. The three wires are carried underground for $2\frac{1}{2}$ miles along the main entry. Clark's insulated wire, No. 2, is used, and branches are carried to the various chambers to operate McMichael's undercutting machines working in the bituminous coal-seam from 6 to 8 ft. wide. The current is used for lighting also, in 16 candle-power incandescent lamps.

New electric coal-cutters are continually being put upon the market,[*] and the improvement in these machines and the introduction of electric pick-machines for hard coal, typifies the rapid advance in the application of electricity to mining work.

In metal mining, electricity has thus far been employed chiefly for pumping and hoisting, though its field will undoubtedly be greatly extended within the next decade.

While it is reported that electric percussion drills are at present in successful use, and exposed to the same hard usage as the air drills,[†] the writer has not yet been fortunate enough to witness such nor to obtain details of their operation underground.

The use of electricity, however, certainly admits of a great saving in the transmission from air-compressor to drills, by the placing of the former underground and closer to the stopes and headings.

That motors are now made that will operate as successfully underground and in damp places, as upon the surface, is conclusively shown by the number of pumps at work in wet shafts to-day. By this it is not meant to say that wet electric motors will operate, but that these are now so insulated and protected that they successfully exclude the moisture of wet shafts and damp foundations.

A most excellent instance of this we have at hand in Superintendent

---

* See "Engineering and Mining Journal," June 16, 1894, p. 559.
† Trans. Am. Inst. M. Engineers, vol. xxiii, p. 405.

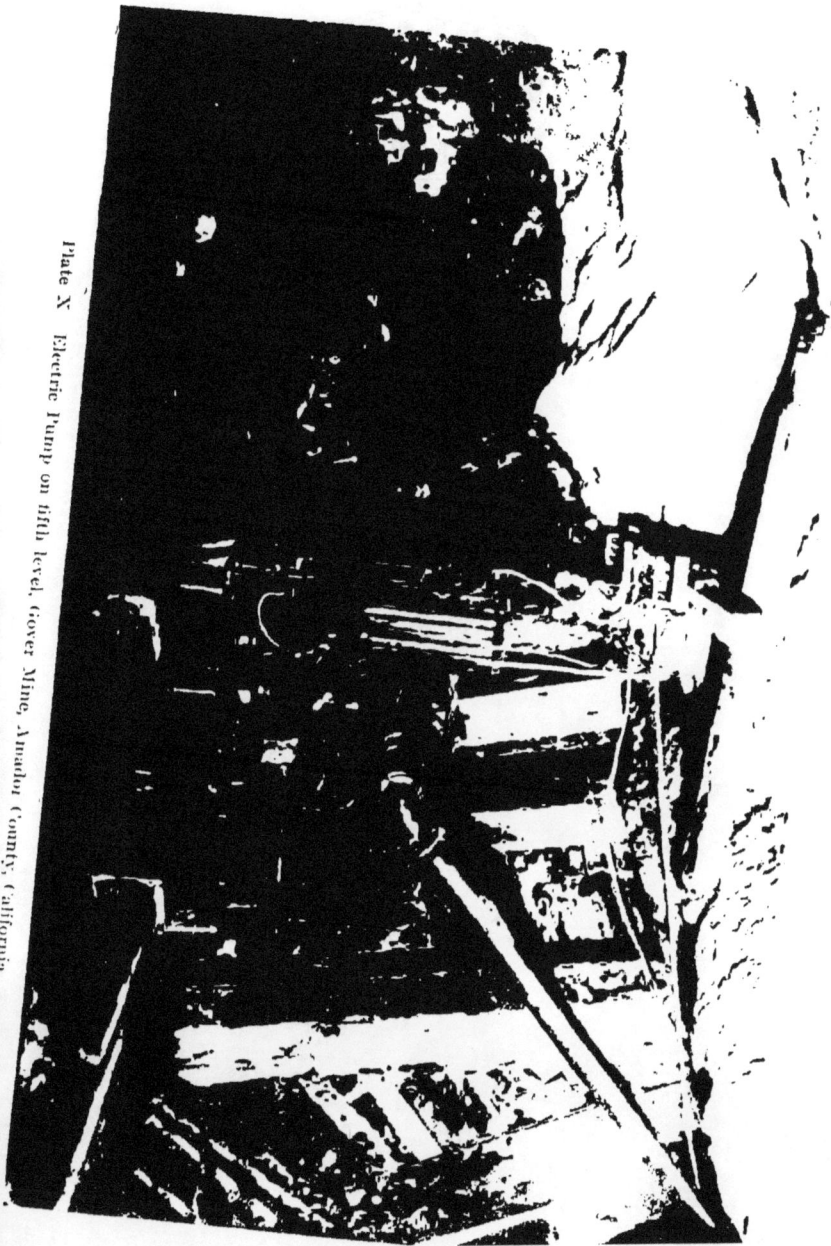

Plate X   Electric Pump on fifth level, Gover Mine, Amador County, California.

Call's letter to the writer, descriptive of the Gover Mining Company's (Amador County, Cal.) plant, as follows:

"Two triple-plunger Dow pumps are used—one with 6 in. plungers raising 12½ miner's inches of water 341 ft. vertically, and one with 5 in. plungers, raising 11 miner's inches of water 208 ft. vertically.

"An Edison dynamo, No. 16, of 50 horse-power capacity, is used, and run at a speed of 820 revolutions per minute, the voltage being 220.

"Sprague motors are used. The one working the larger pump is run at a speed of 1,000 revolutions per minute, giving 20 horse-power; and the other is run at a speed of 1,250 revolutions per minute, producing 15 horse-power. The voltage in the motors is the same as that of the dynamo.

"Copper wire $\frac{5}{16}$ in. in diameter transmits the power from the dynamo to the motors, a distance of about 1,700 ft.—1,000 ft. on the surface, and 700 feet down the shaft. (See Plate X.) The wires cause no trouble whatever in the shaft, retimbering even being done without stopping the pumps. The shaft is quite wet in places.

"The pumps have run three years and four months, pumping during that time 59,000,000 gallons of water.

"The armature of the dynamo burnt out once, owing to injuries received in shipment, the core being shifted. The commutator of the dynamo is turned down about once a year.

"The motors are connected with the pumps by gearing. Rawhide pinions are used on the armature shafts. The rawhide pinions last a year, and are more reliable and more satisfactory than those made of bronze.

"With the motors, the only precaution taken against dampness is a thorough coat of paraffine paint. The smaller motor was run at one time for several hours with the field piece half way under water."

In hoisting, electricity has thus far been applied chiefly to inclines and used in motors of comparatively small capacity or from 10 to 40 horse-power, but its field will undoubtedly be extended to vertical shafts and larger machines as its advantages become more widely recognized.

In the mines of the Aspen Mining and Smelting Company, at Aspen, Colorado, three hoists are in operation underground, each of the two main hoists raising 250 tons up a 60° incline 250 ft. long every twenty-four hours. A 25 horse-power motor in use there raises 3,000 lbs. up a similar slope 275 ft. per minute, and is capable of making the round trip from a depth of 550 ft. in three minutes.

This plant was installed three years ago, and direct-current machines are used throughout. The rapid development of the A. C. systems during the past few years has demonstrated the advantages of the two-phase current for both transmission of the power and the accomplishment of a variety of work at the delivery end of the line.

This is shown in the following description from the "Colliery Guardian" of the recent installation of such a plant at the Decize collieries in France. A noticeable feature, and one already alluded to as a decided advantage of the system, is the ready regulation of varying load on the two currents:

"One of the most interesting cases of the electrical transmission of power for coal-mining purposes in Europe has been completed and set in operation at the Decize collieries, in the Nièvre Department of France, and which are owned by MM. Schneider & Co. This installation is remarkable from the fact that diphase alternating-currents are employed for the transmission, and diphase alternating-current motors are used for reconverting the electrical energy into mechanical power at the different pits. In designing this plant the problem to be solved was to erect a central generating station for the distribution of electrical energy at the different pits where it could be utilized in electro-motors for operating ventilating fans, hauling machinery, pumps, and for lighting purposes. A general idea of what had to be accomplished is shown in the annexed table:

Site.	Distance from Generating Station— Yards.	Electrical Machinery or Lamps Receiving the Current Transmitted.
1.—West.		
Puits des Chagnats	5,090	........30 horse-power electric motor.*
Fendue des Lacets	3,466	........30 horse-power electric motor.*
Puits des Coupes	2,058	........30 horse-power electric motor.*
Puits des Zagots	1,084	.... Electric
2.—Generating Station.		hauling machine of 15 horse-power.†
Various installations		.Six arc and 100 incandescent lamps.‡
3.—East.		
Fendue des Marizy	1,300	....................30 horse-power
Sorting and washing shops of the		electric motor and 24 arc lamps.§
Pré-Charpin	2,490	.... 500 incan-
		descent lamps of 16 candle-power.‡
Champvert	3,250	......,12 horse-power electric motor.‖

* Used for ventilating fan. † Inclined plane. ‡ Lighting. § Ventilating fan and lighting. ‖ Pumping.

"The generating station is situated, respectively, at distances of from 3.1 miles and 1.86 miles from the extreme points which have to be supplied with current. It contains a battery of six boilers and two units (steam engines and dynamos), each of a capacity of 100 kilo-watts; a further unit will shortly be laid down. The two units may be worked singly or in parallel. The engines are of the horizontal non-condensing type, running at 200 revolutions per minute, and driving the diphase alternators by means of belting. A notable feature in this connection is the fact that each electrical unit comprises a twin alternator, or in reality two machines, placed one at each end of the shaft, the driving pulley carrying the engine belt being arranged in the middle of the shaft. Of course, in a case like the present, where current is employed both for lighting and for power purposes, one of the circuits may become more loaded than another, and in this event the equilibrium must be established by varying the ratio of the electro-motive forces. The arrangement adopted in the Decize installation allows of this being accomplished, as each of the two circuits having a distinct field, it is only necessary to vary the exciting current by means of rheostats to get the desired effect. The generators introduced are Zipernowsky ten-pole alternators, with revolving field magnets. The ten-field magnets are connected together in series, and the exciting current is led to them by means of two metallic rings carried on an extension of the driving shaft on the opposite side to that of the driving pulley—that is to say, on an outer extension of the shaft. Two ordinary brass brushes press upon these rings, to which the exciting current is furnished by a direct-current dynamo. This latter machine is operated by a belt from the shaft of the alternator. At 900 revolutions a minute this direct-current dynamo supplies the exciting current for the twin alternator, being between 25 and 30 amperes at 110 volts. The fixed armature of the alternators is formed of ten coils, any one of which can be withdrawn and replaced with little trouble.

"After passing through the switch-board, the current is transmitted mainly by means of overhead wires to the points of utilization, the only portion laid underground being toward the end of the principal line leading to the Chagnats pit. The wires forming the overhead line are of silicon-bronze, and are carried on porcelain insulators attached to poles 24 ft. high. The diameter of the wires constituting the principal line to the western part of the district is 6 mm., and 4 mm. in the case of the remainder of the line. It is noteworthy that the same poles carrying the transmission wires also support telephone wires, the latter being arranged 12 ft. from the ground. In order to counteract the effects of induction in the telephone wires, the line conductors are crossed at distances averaging 540 yds., and by this means the difficulty of understanding conversation along the telephone wires which use the earth as return, has been overcome. The small portion of underground line forms a lead-covered cable, laid in a wooden conduit, as also does the telephone line for the same distance. Suitable lightning conductors are provided at the generating and distributing sub-stations and at intervals along the line. The electro-motors at the sub-stations, where the current is utilized for the different purposes mentioned in the table given above, are of the same type as the generators. These diphase motors are easily set in operation, and are to all intents and purposes left to themselves for several hours together. The only attention they receive is the visit of an employé every six or eight hours to ascertain whether the motors are working properly. The sub-stations are situated in the forest, and the facility of working on this system as compared with the erection in each place of a boiler, steam engine, and ventilating fan, is considered to be remarkable, apart from the question of the cost of transporting fuel."

The advantages of electric power both above and under ground, in point of cleanliness, compactness, ease of transmission, etc., have been so often dwelt upon that it is only surprising they have not been more often availed of by miners everywhere. The next decade will undoubtedly see a wonderful development in the application of electricity to mining operations.

o